NUCLEAR REACTOR THERMAL HYDRAULICS AND OTHER APPLICATIONS

Edited by **Donna Post Guillen**

Nuclear Reactor Thermal Hydraulics and Other Applications
http://dx.doi.org/10.5772/45830
Edited by Donna Post Guillen

Contributors

Alois Hoeld, Weidong Huang, Osama Abd-Elkawi, Ten-See Wang, Sergey Karabasov, Alex Obabko, Paul Fischer, Tim Tautges, Vasily Goloviznin, Mihail Zaitsev, Vladimir Chudanov, Valerii Pervichko, Anna Aksenova, Hernan Tinoco

Published by InTech

Janeza Trdine 9, 51000 Rijeka, Croatia

Publishing Process Manager Iva Simcic
Technical Editor InTech DTP team
Cover InTech Design team

Additional hard copies can be obtained from orders@intechopen.com

Nuclear Reactor Thermal Hydraulics and Other Applications, Edited by Donna Post Guillen
p. cm.
ISBN 978-953-51-0987-7

For EU product safety concerns:
IN TECH d.o.o., Prolaz Marije Krucifikse Kozulić 3, 51000 Rijeka, Croatia,
info@intechopen.com or visit our website at intechopen.com.

Meet the editor

Dr. Donna Post Guillen is in the Energy Systems Integration Department at the Idaho National Laboratory operated for the U.S. Department of Energy. She earned a B.S. in Mechanical Engineering from Rutgers University, an M.S. in Aeronautics from Caltech, and a Ph.D. in Engineering and Applied Science from Idaho State University. Dr. Guillen is a licensed Professional Engineer in the State of Idaho with nearly 30 years of experience in mechanical and systems engineering. The focus of her research is on multiphase computational fluid dynamics and thermal hydraulics for sustainable energy technologies. She has authored or co-authored over 100 technical publications in the form of books, reports, journal articles and conference papers.

Contents

Preface

This book covers a range of thermal hydraulic topics related, but not limited, to nuclear reactors. The purpose is to present research from around the globe that serves to advance our knowledge of nuclear reactor thermal hydraulics and related areas. The focus is on computer code developments and applications to predict fluid flow and heat transfer, with an emphasis on computational fluid dynamic (CFD) methods. This book is divided into two sections. The first section consists of three chapters concerning computational codes and methods applied to nuclear reactor safety. The second section consists of four chapters covering general thermal hydraulic applications.

The overarching theme of the first section of this book is thermal hydraulic models and codes to address safety behaviour of nuclear power plants. Accurate predictions of heat transfer and fluid flow are required to ensure effective heat removal under all conditions. The section begins with a chapter discussing the theoretical development of thermal-hydraulic approaches to coolant channel analysis. These traditional methods are widely used in system codes to evaluate nuclear power plant performance and safety. The second chapter examines several fully unsteady computational models in the framework of large eddy simulations implemented for a thermal hydraulic transport problem relevant to the design of nuclear power plant piping systems. A comparison of experimental data from a classic benchmark problem with the numerical results from three simulation codes is given. The third chapter addresses the issue of properly modeling thermal mixing in Light Water Nuclear Reactors. A CFD approach is advocated, which allows the flow structures to develop and properly capture the mixing properties of turbulence.

The second section of this book includes chapters focusing on the application of CFD to crosscutting thermal hydraulic phenomena. In line with best practices for CFD, the simulations are supported by relevant experimental data. The section begins with a chapter describing a thermal hydraulic design and analysis methodology for a nuclear thermal propulsion development effort. Modern computational fluid dynamics and heat transfer methods are used to predict thermal, fluid, and hydrogen environments of a hypothetical solid-core, nuclear thermal engine designed in the 1960s. The second chapter in this section investigates the applicability of several CFD approaches to modeling mixing and agitation in a stirred tank reactor. The results are compared with experimentally-obtained velocity and turbulence parameters to determine the most appropriate methodology. The third chapter in this section presents the results of CFD simulations used to study the hydrodynamics and heat transfer processes in a two-dimensional gas fluidized bed. The final chapter uses CFD to predict the thermal hydraulics surrounding the design of a spallation target system for an Accelerator Driven System.

Our ability to simulate larger problems with greater fidelity has vastly expanded over the past decade. The collection of material presented in this book is but a small contribution to the important topic of thermal hydraulics. The contents of this book will interest researchers, scientists, engineers and graduate students.

Dr. Donna Post Guillen
Group Lead, Advanced Process and Decision Systems Department,
Idaho National Laboratory, USA

Section 1

CFD Applications for Nuclear Reactor Safety

Chapter 1

The Coolant Channel Module CCM — A Basic Element for the Construction of Thermal-Hydraulic Models and Codes

Alois Hoeld

Additional information is available at the end of the chapter

http://dx.doi.org/10.5772/53372

1. Introduction

The development of LWR Nuclear Power Plants (NPP) and the question after their safety behaviour have enhanced the need for adequate efficient theoretical descriptions of these plants. Thus thermal-hydraulic models and, based on them, effective computer codes played already very early an important role within the field of NPP safety research. Their objective is to describe both the steady state and transient behaviour of characteristic key parameters of a single- or two-phase fluid flowing along corresponding loops of such a plant and thus also along any type of heated or non-heated coolant channels being a part of these loops in an adequate way.

Due to the presence of discontinuities in the first principle of mass conservation of a two-phase flow model, caused at the transition from single- to two-phase flow and vice versa, it turned out that the direct solution of the basic conservation equations for mixture fluid along such a coolant channel gets very complicated. Obviously many discussions have and will continue to take place among experts as to which type of theoretical approach should be chosen for the correct description of thermal-hydraulic two-phase problems when looking at the wide range of applications. What is thus the most appropriate way to deal with such a special thermal-hydraulic problem?

With the introduction of a 'Separate-Phase Model Concept' already very early an efficient way has been found how to circumvent these upcoming difficulties. Thereby a solution method has been proposed with the intention to separate the two-phases of such a mixture-flow in parts of the basic equations or even completely from each other. This yields a system of 4-, 5- or sometimes even 6-equations by splitting each of the conservation equations into two so-called 'field equations'. Hence, compared to the four independent parameters characterising the mixture fluid, the separate-phase systems demand a much higher number of additional variables and special assumptions. This has the additional consequence that a number of speculative relations had to be incorporated into the theoretical description of such a module and an enormous amount of CPU-time has to be expended for the solution of the resulting sets of differential and analytical equations in a computer code. It is also clear that, based on such assumptions, the interfacial relations both between the (heated or cooled) wall but also between each of the two phases are completely rearranged. This raises the difficult question of how to describe in a realistic way the direct heat input into and between the phases and the movement resp. the friction of the phases between them. In such an approach this problem is solved by introducing corresponding exchange (=closure) terms between the equations based on special transfer (= closure) laws. Since they can, however, not be based on fundamental laws or at least on experimental measurements this approach requires a significant effort to find a correct formulation of the exchange terms between the phases. It must therefore be recognised that the quality of these basic equations (and especially their boundary conditions) will be intimately related to the (rather artificial and possibly speculative) assumptions adopted if comparing them with the original conservation laws of the 3-equation system and their constitutive equations as well. The problem of a correct description of the interfacial reaction between the phases and the wall remains. Hence, very often no consistency between different separate-phase models due to their underlying assumptions can be stated. Another problem arises from the fact that special methods have to be foreseen to describe the moving boiling boundary or mixture level (or at least to estimate their 'condensed' levels) in such a mixture fluid (see, for example, the 'Level Tracking' method in TRAC). Additionally, these methods show often deficiencies in describing extreme situations such as the treatment of single- and two-phase flow at the ceasing of natural circulation, the power situations if decreasing to zero etc. The codes are sometimes very inflexible, especially if they have to provide to a very complex physical system also elements which belong not to the usual class of 'thermal-hydraulic coolant channels'. These can, for example, be nuclear kinetic considerations, heat transfer out of a fuel rod or through a tube wall, pressure build-up within a compartment, time delay during the movement of an enthalpy front along a downcomer, natural circulation along a closed loop, parallel channels, inner loops etc.

However, despite of these difficulties the 'Separate-Phase Models' have become increasingly fashionable and dominant in the last decades of thermal-hydraulics as demonstrated by the widely-used codes TRAC (Lilles et al.,1988, US-NRC, 2001a), CATHENA (Hanna, 1998), RELAP (US-NRC,2001b, Shultz,2003), CATHARE (Bestion,1990), ATHLET (Austregesilo et al., 2003, Lerchl et al., 2009).

Within the scope of reactor safety research very early activities at the Gesellschaft für Anlagen- und Reaktorsicherheit (GRS) at Garching/Munich have been started too, developing thermal-hydraulic models and digital codes which could have the potential to describe in a detailed way the overall transient and accidental behaviour of fluids flowing along a reactor core but also the main components of different Nuclear Power Plant (NPP) types. For one of these components, namely the natural circulation U-tube steam generator together with its feedwa-

http://dx.doi.org/10.5772/53372

ter and main steam system, an own theoretical model has been derived. The resulting digital code UTSG could be used both in a stand-alone way but also as part of more comprehensive transient codes, such as the thermal-hydraulic GRS system code ATHLET. Together with a high level simulation language GCSM (General Control Simulation Module) it could be taken care of a manifold of balance-of-plant (BOP) actions too. Based on the experience of many years of application both at the GRS and a number of other institutes in different countries but also due to the rising demands coming from the safety-related research studies this UTSG theory and code has been continuously extended, yielding finally a very satisfactory and mature code version UTSG-2.

During the research work for the development of an enhanced version of the code UTSG-2 it arose finally the idea to establish an own basic element which is able to simulate the thermal-hydraulic mixture-fluid situation within any type of cooled or heated channel in an as general as possible way. It should have the aim to be applicable for any modular construction of complex thermal-hydraulic assemblies of pipes and junctions. Thereby, in contrast to the above mentioned class of 'separate-phase' modular codes, instead of separating the phases of a mixture fluid within the entire coolant channel an alternative theoretical approach has been proposed, differing both in its form of application but also in its theoretical background. To circumvent the above mentioned difficulties due to discontinuities resulting from the spatial discretization of a coolant channel, resulting eventually in nodes where a transition from single- to two-phase flow and vice versa can take place, a special and unique concept has been proposed. Thereby it has been assumed that each coolant channel can be seen as a (basic) channel (BC) which can, according to their different flow regimes, be subdivided into a number of sub-channels (SC-s). It is clear that each of these SC-s can consist of only two types of flow regimes. A SC with just a single-phase fluid, containing exclusively either sub-cooled water, superheated steam or supercritical fluid, or a SC with a two-phase mixture. The theoretical considerations of this 'Separate-Region Approach' can then (within the class of mixture-fluid models) be restricted to only these two regimes. Hence, for each SC type, the 'classical' 3 conservation equations for mass, energy and momentum can be treated in a direct way. In case of a sub-channel with mixture flow these basic equations had to be supported by a drift flux correlation (which can take care also of stagnant or counter-current flow situations), yielding an additional relation for the appearing fourth variable, namely the steam mass flow.

The main problem of the application of such an approach lies in the fact that now also varying SC entrance and outlet boundaries (marking the time-varying phase boundary positions) have to be considered with the additional difficulty that along a channel such a SC can even disappear or be created anew. This means that after an appropriate nodalization of such a BC (and thus also it's SC-s) a 'modified finite volume method' (among others based on the Leibniz Integration Rule) had to be derived for the spatial discretization of the fundamental partial differential equations (PDE-s) which represent the basic conservation equations of thermal-hydraulics for each SC. Furthermore, to link within this procedure the resulting mean nodal with their nodal boundary function values an adequate quadratic polygon approximation method (PAX) had to be established. The procedure should yield finally for each SC type (and thus also the complete BC) a set of non-linear ordinary differential equations of 1st order (ODE-s).

It has to be noted that besides the suggestion to separate a (basic) channel into regions of different flow types this special PAX method represents, together with the very thoroughly tested packages for drift flux and single- and two-phase friction factors, the central part of the here presented 'Separate - Region Approach'. An adequate way to solve this essential problem could be found and a corresponding procedure established. As a result of these theoretical considerations an universally applicable 1D thermal-hydraulic drift-flux based separate-region coolant channel model and module CCM could be established. This module allows to calculate automatically the steady state and transient behaviour of the main characteristic parameters of a single- and two-phase fluid flowing within the entire coolant channel. It represents thus a valuable tool for the establishment of complex thermal-hydraulic computer codes. Even in the case of complicated single- and mixture fluid systems consisting of a number of different types of (basic) coolant channels an overall set of equations by determining automatically the nodal non-linear differential and corresponding constitutive equations needed for each of these sub- and thus basic channels can be presented. This direct method can thus be seen as a real counterpart to the currently preferred and dominant 'separate-phase models'.

To check the performance and validity of the code package CCM and to verify it the digital code UTSG-2 has been extended to a new and advanced version, called UTSG-3. It has been based, similarly as in the previous code UTSG-2, on the same U-tube, main steam and downcomer (with feedwater injection) system layout, but now, among other essential improvements, the three characteristic channel elements of the code UTSG-2 (i.e. the primary and secondary side of the heat exchange region and the riser region) have been replaced by adequate CCM modules.

It is obvious that such a theoretical 'separate-region' approach can disclose a new way in describing thermal-hydraulic problems. The resulting 'mixture-fluid' technique can be regarded as a very appropriate way to circumvent the uncertainties apparent from the separation of the phases in a mixture flow. The starting equations are the direct consequence of the original fundamental physical laws for the conservation of mass, energy and momentum, supported by well-tested heat transfer and single- and two-phase friction correlation packages (and thus avoiding also the sometimes very speculative derivation of the 'closure' terms). In a very comprehensive study by (Hoeld, 2004b) a variety of arguments for the here presented type of approach is given, some of which will be discussed in the conclusions of chapter 6.

The very successful application of the code combination UTSG-3/CCM demonstrates the ability to find an exact and direct solution for the basic equations of a 'non-homogeneous drift-flux based thermal-hydraulic mixture-fluid coolant channel model'. The theoretical background of CCM will be described in very detail in the following chapters.

http://dx.doi.org/10.5772/53372

For the establishment of the corresponding (digital) module CCM, based on this theoretical model very specific methods had to be achieved. Thereby the following points had to be taken into account:

- The code has to be easily applicable, demanding only a limited amount of directly available input data. It should make it possible to simulate the thermal-hydraulic mixture-fluid situation along any cooled or heated channel in an as general as possible way and thus describe any modular construction of complex thermal-hydraulic assemblies of pipes and junctions. Such an universally applicable tool can then be taken for calculating the steady state and transient behaviour of all the characteristic parameters of each of the appearing coolant channels and thus be a valuable element for the construction of complex computer codes. It should yield as output all the necessary time-derivatives and constitutive parameters of the coolant channels required for the establishment of an overall thermal-hydraulic code.

- It was the intention of CCM that it should act as a complete system in its own right, requiring only BC (and not SC) related, and thus easily available input parameters (geometry data, initial and boundary conditions, parameters resulting from the integration etc.). The partitioning of BC-s into SC-s is done at the beginning of each recursion or time-step automatically within CCM, so no special actions are required of the user.

- The quality of such a model is very much dependent on the method by which the problem of the varying SC entrance and outlet boundaries can be solved. Especially if they cross BC node boundaries during their movement along a channel. For this purpose a special 'modified finite element-method' has been developed which takes advantage of the 'Leibniz' rule for integration (see eq.(15)).

- For the support of the nodalized differential equations along different SC-s a 'quadratic polygon approximation' procedure (PAX) was constructed in order to interrelate the mean nodal with the nodal boundary functions. Additionally, due to the possibility of varying SC entrance and outlet boundaries, nodal entrance gradients are required too (See section 3.3).

- Several correlation packages such as, for example, packages for the thermodynamic properties of water and steam, heat transfer coefficients, drift flux correlations and single- and two-phase friction coefficients had to be developed and implemented (See sections 2.2.1 to 2.2.4).

- Knowing the characteristic parameters at all SC nodes (within a BC) then the single- and two-phase parameters at all node boundaries of the entire BC can be determined. And also the corresponding time-derivatives of the characteristic averaged parameters of coolant temperatures resp. void fraction over these nodes. This yields a final set of ODE-s and constitutive equations.

- In order to be able to describe also thermodynamic non-equilibrium situations it can be assumed that each phase is represented by an own with each other interacting BC. For these purpose in the model the possibility of a variable cross flow area along the entire channel had to be considered as well.

- Within the CCM procedure two further aspects play an important role. These are, however, not essential for the development of mixture-fluid models but can help enormously to enhance the computational speed and applicability of the resulting code when simulating a complex net of coolant pipes:
- The solution of the energy and mass balance equations at each intermediate time step will be performed independently from momentum balance considerations. Hence the heavy CPU-time consuming solution of stiff equations can be avoided (Section 3.6).
- This decoupling allows then also the introduction of an 'open' and 'closed channel' concept (see section 3.11). Such a special method can be very helpful in describing complex physical systems with eventually inner loops. As an example the simulation of a 3D compartment by parallel channels can be named (Jewer et al., 2005).

The application of a direct mixture-fluid technique follows a long tradition of research efforts. Ishii (1990), a pioneer of two-fluid modelling, states with respect to the application of effective drift-flux correlation packages in thermal-hydraulic models: 'In view of the limited data base presently available and difficulties associated with detailed measurements in two-phase flow, an advanced mixture-fluid model is probably the most reliable and accurate tool for standard two-phase flow problems'. There is no new knowledge available to indicate that this view is invalid.

Generally, the mixture-fluid approach is in line with (Fabic, 1996) who names three strong points arguing in favour of this type of drift-flux based mixture-fluid models:

- They are supported by a wealth of test data,
- they do not require unknown or untested closure relations concerning mass, energy and momentum exchange between phases (thus influencing the reliability of the codes),
- they are much simpler to apply,

and, it can be added,

- discontinuities during phase changes can be avoided by deriving special solution procedures for the simulation of the movement of these phase boundaries,
- the possibility to circumvent a set of 'stiff' ODE-s saves an enormous amount of CPU time which means that the other parts of the code can be treated in much more detail.

A documentation of the theoretical background of CCM will be given in very condensed form in the different chapters of this article. For the establishment of the corresponding (digital) module CCM, based on this theoretical model, very specific methods had to be achieved.

The here presented article is an advanced and very condensed version of a paper being already published in a first Open Access Book of this INTECH series (Hoeld, 2011a). It is updated to the newest status in this field of research. An example for an application of this module within the UTSG-3 steam generator code is given in (Hoeld 2011b).

http://dx.doi.org/10.5772/53372

2. Thermal-hydraulic drift-flux based mixture fluid approach

2.1. Thermal-hydraulic conservation equations

Thermal-hydraulic single-phase or mixture-fluid models for coolant channels or, as presented here, for each of the sub-channels are generally based on a number of fundamental physical laws, i.e., they obey genuine conservation equations for mass, energy and momentum. And they are supported by adequate constitutive equations (packages for thermo-dynamic and transport properties of water and steam, for heat transfer coefficients, for drift flux, for single- and two-phase friction coefficients etc.).

In view of possible applications as an element in complex thermal-hydraulic ensembles outside of CCM eventually a fourth and fifth conservation law has to be considered too. The fourth law, namely the volume balance, allows then to calculate the transient behaviour of the overall absolute system pressure. Together with the local pressure differences then the absolute pressure profile along the BC can be determined. The fifth physical law is based on the (trivial) fact that the sum of all pressure decrease terms along a closed loop must be zero. This is the basis for the treatment of the thermal-hydraulics of a channel according to a 'closed channel concept' (See section 3.11). It refers to one of the channels within the closed loop where the BC entrance and outlet pressure terms have to be assumed to be fixed. Due to this concept then the necessary entrance mass flow term has be determined in order to fulfil the demand from momentum balance.

2.1.1. *Mass balance (Single- and two-phase flow)*

$$\frac{\partial}{\partial t}\{A[(1-\alpha)r_W+\alpha r_S]\}+\frac{\partial}{\partial z}G=0 \tag{1}$$

Containing the density terms ϱ_W and ϱ_S for sub-cooled or saturated water and saturated or superheated steam, the void fraction α and the cross flow area A which can eventually be changing along the coolant channel. It determines, after a nodalization, the total mass flow $G=G_W+G_S$ at each node outlet in dependence of its node entrance value.

2.1.2. *Energy balance (Single- and two-phase flow)*

$$\frac{\partial}{\partial t}\{A[(1-\alpha)r_W h_W+\alpha r_S h_S-P]\}+\frac{\partial}{\partial z}\left[G_W h_W+G_S h_S\right]=q_{TWL}=U_{TW}q_{TWF}=A\ q_D \tag{2}$$

Containing the enthalpy terms h_W and h_S for sub-cooled or saturated water and saturated or superheated steam. As boundary values either the 'linear power q_{TWL}', the 'heat flux q_{TWF}' along the heated (or cooled) tube wall (with its heated perimeter U_{TW}) or the local 'power density term q_D' (being transferred into the coolant channel with its cross section A) are demanded to be known (See also sections 2.2.4 and 3.5). They are assumed to be directed into the coolant (then having a positive sign).

2.1.3. Momentum balance (Single- and two-phase flow)

$$\frac{\partial}{\partial t}(G_F)+(\frac{\partial P}{\partial z})=(\frac{\partial P}{\partial z})_A+(\frac{\partial P}{\partial z})_S+(\frac{\partial P}{\partial z})_F+(\frac{\partial P}{\partial z})_X \tag{3}$$

describing either the pressure differences (at steady state) or (in the transient case) the change in the total mass flux (G_F =G/A) along a channel (See chapter 3.10).

The general pressure gradient ($\frac{\partial P}{\partial z}$) can be determined in dependence of

- the mass acceleration

$$(\frac{\partial P}{\partial z})_A=-\frac{\partial}{\partial z}\left[\left(G_{FW}v_W+G_{FS}v_S\right)\right] \tag{4}$$

with v_S and v_W denoting steam and water velocities given by the eqs.(19) and (20),

- the static head

$$(\frac{\partial P}{\partial z})S=-\cos(\Phi_{ZG})\,g_C[\alpha\rho_S+(1-\alpha)\rho W] \tag{5}$$

with Φ_{ZG} representing the angle between z-axis and flow direction. Hence

$\cos(\Phi_{ZG})=\pm\ \Delta z_{EL}/\Delta z_L$, with Δz_L denoting the nodal length and Δz_{EL} the nodal elevation height (having a positive sign at upwards flow).

- the single- and/or two-phase friction term

$$(\frac{\partial P}{\partial z})_F=-f_R\frac{G_F|G_F|}{2\,d_{HW}\,\rho} \tag{6}$$

with a friction factor derived from corresponding constitutive equations (section 2.2.2)

and finally

- the direct perturbations $(\partial P/\partial z)_X$ from outside, arising either by starting an external pump or considering a pressure adjustment due to mass exchange between parallel channels.

2.2. Constitutive equations

For the exact description of the steady state and the transient behaviour of single- or two-phase fluids a number of mostly empirical constitutive correlations are, besides the above mentioned conservation equations, demanded. To bring a structure into the manifold of existing correlations established by various authors, to find the best fitting correlations for the different fields of application and to get a smooth transfer from one to another of them special and effective correlation packages had to be developed. Their validities can be and has been tested out-of-pile by means of adequate driver codes. Obviously, my means of this method improved correlations can easily be incorporated into the existing theory.

A short characterization of the main packages being applied within CCM is given below. For more details see (Hoeld, 2011a).

2.2.1. *Thermodynamic and transport properties of water and steam*

The different thermodynamic properties for water and steam (together with their derivatives with respect to P and T, but also P and h) demanded by the conservation and constitutive equations have to be determined by applying adequate water/steam tables. This is, for light-water systems, realized in the code package MPP (Hoeld, 1996 and 2011a).

Then the time-derivatives of these thermodynamic properties which respect to their independent local parameters (for example of an enthalpy term h) can be represented as

$$\frac{d}{dt}h(z,t) = \frac{d}{dt}h\left[T(z,t),P(z,t)\right] = h^T \frac{d}{dt}T_{Mn}(z,t) + h^P \frac{d}{dt}P_{Mn}(z,t) \qquad (7)$$

Additionally, corresponding thermodynamic transport properties such as 'dynamic viscosity' and 'thermal heat conductivity' (and thus the 'Prantl number') are asked from some constitutive equations too as this can be stated, for example, for the code packages MPPWS and MPPETA (Hoeld, 1996). All of them have been derived on the basis of tables given by (Schmidt and Grigull,1982) and (Haar et al., 1988).

Obviously, the CCM method is also applicable for other coolant systems (heavy water, gas) if adequate thermodynamic tables for this type of fluids are available.

2.2.2. *Single and two-phase friction factors*

The friction factor f_R needed in eq.(6) can in case of single-phase flow be set, as proposed by (Moody, 1994), equal to the Darcy-Weisbach single-phase friction factor.

The corresponding coefficient for two-phase flow has to be extended by means of a two-phase multiplier Φ^2_{2PF} as recommended by (Martinelli-Nelson, 1948).

For more details see again (Hoeld, 2011a).

2.2.3. *Drift flux correlation*

Usually, the three conservation equations (1), (2) and (3) demand for single-phase flow the three parameters G, P and T as independent variables. In case of two-phase flow, they are, however, dependent on four of them, namely G, P, α and G_S. This means, the set has to be completed by an additional relation. This can be achieved by any two-phase correlation, acting thereby as a 'bridge' between G_S and α. For example, by a slip correlation. However, to take care of stagnant or counter-current flow situations too an effective drift-flux correlation seemed here to be more appropriate. For this purpose an own package has been established, named MDS (Hoeld, 2001 and 2002a). Due to the different requirements in the application of CCM it turned out that it has a number of advantages if choosing the 'flooding-based full-range' Sonnenburg correlation (Sonnenburg, 1989) as basis for MDS. This correlation combines the common drift-flux procedure being formulated by (Zuber-Findlay, 1965) and expanded by (Ishii-Mishima, 1980) and (Ishii, 1990) etc. with the modern envelope theory. The correlation in the final package MDS had, however, to be rearranged in such a way that also the special cases of $\alpha \rightarrow 0$ or $\alpha \rightarrow 1$ are included and that, besides their absolute values and corresponding slopes, also the gradients of the approximation function can be made available for CCM. Additionally, an inverse form had to be installed (needed, for example, for the steady state conditions) and, eventually, also considerations with respect to possible entrainment effects be taken care.

For the case of a vertical channel this correlation can be represented as

$$v_D = 1.5\, v_{WLIM} C_0 C_{VD} [(1+C_{VD}{}^2)^{3/2} - (1.5+C_{VD}{}^2) C_{VD}] \text{ with } v_D \rightarrow v_{D0} = \frac{9}{16} C_0 v_{WLIM} \text{ if } \alpha \rightarrow 0 \tag{8}$$

where the coefficient C_{VD} is given by

$$C_{VD} = \frac{2}{3} \frac{v_{SLIM}}{v_{WLIM}} \frac{1-C_0\alpha}{C_0\alpha} \tag{9}$$

with (in case of a heated or non-heated channel) $C_0 \rightarrow 0$ or 1 if $\alpha \rightarrow 0$ resp. $C_0 \rightarrow 1$ if $\alpha \rightarrow 1$ and corresponding drift velocity terms v_D according to eq.(8.)

The resulting package MDS yields in combination with an adequate correlation for the phase distribution parameter C_0 relations for the limit velocities v_{SLIM} and v_{WLIM} and thus (independently of the total mass flow G which is important for the theory below) relations for the drift velocity v_D with respect to the void fraction α. All of them are dependent on the given 'system pressure P', the 'hydraulic diameter d_{HY}' (with respect to the wetted surface A_{TW} and its inclination angle Φ_{ZG}), on specifications about the geometry type (L_{GTYPE}) and, for low void fractions, the information whether the channel is heated or not.

The drift flux theory can thus be expressed in dependence of a (now already on G dependent) steam mass flow (or flux) term

$$G_S = \frac{\rho^{//}}{\rho^{/}} \frac{\alpha}{C_{GC}} (C_0 G + A\rho^{/} v_D) = A G_{FS} \frac{\alpha}{C_{GC}} \tag{10}$$

with the coefficient

$$C_{GC}= 1-(1-\frac{\rho^{//}}{\rho^{/}})\alpha C_0 \rightarrow 1 \text{ if } \alpha \rightarrow \frac{\rho^{//}}{\rho^{/}} \text{ if } \alpha \rightarrow 1 \tag{11}$$

Knowing now the fourth variable then, by starting from their definition equations, relations for all the other characteristic two-phase parameters can be established. Such two-phase parameters could be the 'phase distribution parameter C_0', the 'water and steam mass flows G_W and G_S', 'drift, water, steam and relative velocities v_D, v_W,v_S and v_R' (with special values for v_S if α -> 0 and v_W if α -> 1) and eventually the 'steam quality X'. Their interrelations are shown, for example, in the tables of (Hoeld, 2001 and 2002a). Especially the determination of the steam mass flow gradient

$$\begin{aligned} G_S^{\alpha}) \rightarrow G_{S0}^{\alpha} &= \frac{\rho^{//}}{\rho^{/}}(C_{00}G+Ar^{/}v_{D0}) = Ar^{//}v_{S0} \quad \text{or} =0 \quad \text{if } \alpha \rightarrow 0 \text{ and } L_{HEATD}= 0 \text{ or } 1 \\ \rightarrow G_{S1}^{\alpha} &= A\frac{\rho^{//}}{\rho^{/}}(1+C_{01}{}^{(\alpha)})\,(G - r^{/}\,{}^{/}v_{SLIM}) = Ar^{/}v_{W1} \quad \text{if } \alpha \rightarrow 1 \end{aligned} \tag{12}$$

will play (as shown, for example, in eq.(52)) an important part, if looking to the special situation that the entrance or outlet position of a SC is crossing a BC node boundary ($\alpha \rightarrow$ 0 or $\rightarrow$ 1). This possibility makes the drift-flux package MDS to an indispensable part in the nodalization procedure of the mixture-fluid mass and energy balance.

At a steady state situation as a result of the solution of the basic (algebraic) set of equations the steam mass flow term G_S acts as an independent variable (and not the void fraction α). The same is the case after an injection of a two-phase mixture coming from a 'porous' channel or an abrupt change in steam mass flux G_{FS}, as this can take place after a change in total mass flow or in the cross flow area at the entrance of a following BC. Then the total and the steam mass flow terms G and G_S have to be taken as the basis for further two-phase considerations. The void fraction α and other two-phase parameters (v_D, C_0) can now be determined from an inverse (INV) form of this drift-flux correlation (with G_S now as input).

2.2.4. Heat transfer coefficients

The nodal BC heat power terms Q_{BMk} into the coolant are needed (as explained in section 3.5) as boundary condition for the energy balance equation (2). If they are not directly available (as this is the case for electrically heated loops) they have to be determined by solving an adequate Fourier heat conduction equation, demanding as boundary condition

$$q_F=\alpha_{TW}(T_{TW}-T) = \frac{q_{LTW}}{U_{TW}} = \frac{A}{U_{TW}} q_D \tag{13}$$

Such a procedure is, for example, presented in (Hoeld, 2002b, 2011) for the case of heat conduction through a U-tube wall (See also section 3.5).

For this purpose adequate heat transfer coefficients are demanded. This means a method had to be found for getting these coefficients α_{TW} along a coolant channel at different flow regimes. In connection with the development of the UTSG code (and thus also of CCM) an own very comprehensive heat transfer coefficient package, called HETRAC (Hoeld 1988a), has been established.

This classic method is different to the 'separate-phase' models where it has to be taken into account that the heat is transferred both directly from the wall to each of the two possible phases but also exchanged between them. There arises then the question how the corresponding heat transfer coefficients for each phase should look like.

3. Coolant channel module CCM

3.1. Channel geometry and finite-difference nodalization

The theoretical considerations take advantage of the fact that, as sketched in fig.1, a 'basic' coolant channel (BC) can, as already pointed-out, according to their flow regimes (characterized by the logical $L_{FTYPE} = 0, 1, 2$ or 3) be subdivided into a number (N_{SCT}) of sub-channels (SC-s), each distinguished by their characteristic key numbers (N_{SC}). Obviously, it has to be taken into account that their entrance and outlet SC-s can now have variable entrance and/or outlet positions.

The entire BC, with its total length $z_{BT} = z_{BA}-z_{BE}$, can then, for discretization purposes, be also subdivided into a number of (not necessarily equidistant) N_{BT} nodes. Their nodal positions are z_{BE}, z_{Bk} (with k=1,N_{BT}), the elevation heights z_{ELBE}, z_{ELk}, the nodal length $\Delta z_{Bk}=z_{Bk}-z_{Bk-1}$, the nodal elevations $\Delta z_{ELBk}=z_{ELBk}-z_{ELBk-1}$, with eventually also locally varying cross flow and average areas A_{Bk} and $A_{BMk}=0.5(A_{Bk}+A_{Bk-1})$ and their slopes $A^{z}_{Bk} = (A_{Bk}-A_{Bk-1})/\Delta z_{Bk}$, a hydraulic diameter d_{HYBk} and corresponding nodal volumes $V_{BMk} = \Delta z_{Bk} A_{BMk}$. All of them can be assumed to be known from input.

As a consequence, each of the sub-channels (SC-s) is then subdivided too, now into a number of N_{CT} SC nodes with geometry data being identical to the corresponding BC values, except, of course, at their entrance and outlet positions. The SC entrance position z_{CE} and their function f_{CE} are either identical with the BC entrance values z_{BE} and f_{BE} or equal to the outlet values of the SC before. The SC outlet position (z_{CA}) is either limited by the BC outlet (z_{BA}) or characterized by the fact that the corresponding outlet function has reached an upper or lower limit (f_{LIMCA}). This the term represents either a function at the boiling boundary, a mixture level or the start position of a supercritical flow. Such a function follows from the given BC limit values and will, in the case of single-phase flow, be equal to the saturation temperature T_{SATCA} or saturation enthalpies ($h^{/}$ or $h^{//}$ if L_{FTYPE}=1 or 2). In the case of two-phase flow (L_{FTYPE}=0) it has to be equal to a void fraction of $\alpha = 1$ or $= 0$. The moving SC inlet and outlet positions z_{CE} and z_{CA} can (together with their corresponding BC nodes N_{BCE} and $N_{BCA} = N_{BCE}+N_{CT}$) be determined according to the conditions ($z_{BNk-1} \leq z_{CE} < z_{BNk}$ at $k = N_{BCE}$) and ($z_{BNk-1} \leq z_{CA} < z_{BNk}$ at $k = N_{BCA}$). Then also the total number of SC nodes ($N_{CT}=N_{BCA}-N_{BCE}$) is given, the connection between n and k (n=k-N_{BCE} with n=1, N_{CT}), the corresponding positions (z_{Nn}, z_{ELCE}, z_{ELNn}), their lengths ($\Delta z_{Nn}=z_{Nn}-z_{Nn-1}$), elevations ($\Delta z_{ELNn}=z_{ELNn}-z_{ELNn-1}$) and volumes ($V_{Mn}=z_{Nn}A_{Mn}$) and nodal boundary and mean nodal flow areas (A_{Nn}, A_{Mn}).

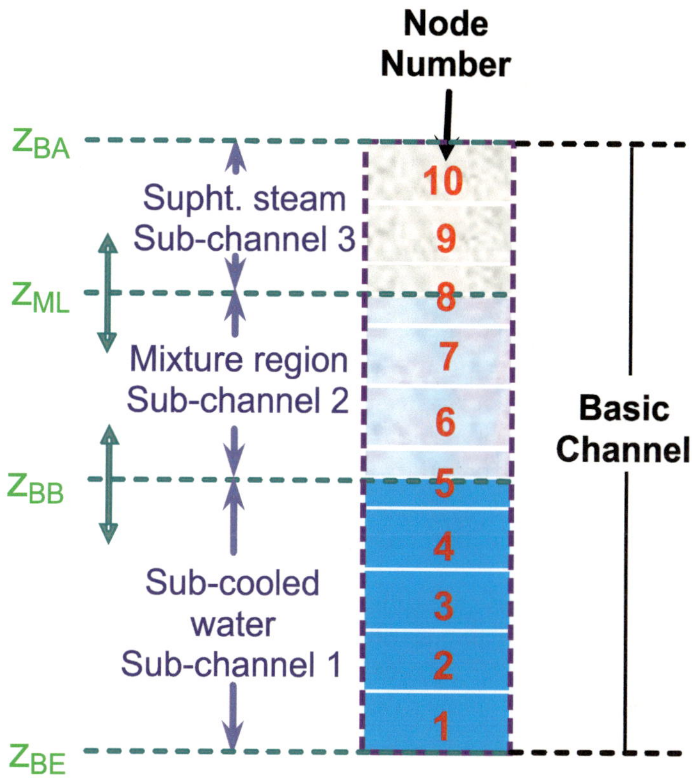

Figure 1. Subdivision of a 'basic channel (BC)' into 'sub-channels (SC-s)' according to their flow regimes and their discretization

3.2. Spatial discretization of PDE-s of 1-st order (Modified finite element method)

Based on this nodalization the spatial discretization of the fundamental eqs.(1) to (3) can be performed by means of a 'modified finite element method'. This means that if a partial differential equation (PDE) of 1-st order having the general form with respect to a general solution function f(z,t)

$$\frac{\partial}{\partial t}f(z,t)+\frac{\partial}{\partial z}H\left[f(z,t)\right]=R\left[f(z,t)\right] \qquad (14)$$

is integrated over the length of a SC node three types of discretization elements can be expected:

- Integrating a function f(z,t) over a SC node n yields the nodal mean function values f_{Mn},
- integrating over the gradient of a function f(z,t) yields a difference of functions values ($f_{Nn} - f_{Nn-1}$) at their node boundaries

and, finally,

- integrating over a time-derivative of a function (by applying the 'Leibniz' rule) yields

$$\int_{z_{Nn-1}(t)}^{z_{Nn}(t)} \frac{\partial}{\partial t} f(z,t)dz = \Delta z_{Nn}(t)\frac{d}{dt}f_{Mn}(t) - [f_{Nn}(t) - f_{Mn}(t)]\frac{d}{dt}z_{Nn}(t)$$
$$- [f_{Mn}(t) - f_{Nn-1}(t)]\frac{d}{dt}z_{Nn-1}(t) \qquad (n=1, N_{CT}) \tag{15}$$

This last rule plays for the here presented 'separate-region mixture-fluid approach' an outstanding part. It allows (together with PAX) to determine in a direct way the time-derivatives of parameters which represent either a boiling boundary, mixture or a supercritical level. This procedure differs considerably from some of the 'separate-phase methods' where, as already pointed out, very often only the collapsed levels of a mixture fluid can be calculated.

3.3. Quadratic polygon approximation procedure PAX

According to the above described three different types of possible discretization elements the solution of the set of algebraic equations will in the steady state case (as shown later-on) yield function values (f_{Nn}) at the node boundaries (z_{Nn}), the also needed mean nodal functions (f_{Mn}) will then have to be determined on the basis of f_{Nn}. On the other hand, the solution of the set of ordinary differential equations will in the transient case now yield the mean nodal functions f_{Mn} as a result, the also needed nodal boundary values f_{Nn} will have to be estimated on the basis of f_{Mn}.

It is thus obvious that appropriate methods had to be developed which can help to establish relations between such mean nodal (f_{Mn}) and node boundary (f_{Nn}) function values. Different to the 'separate-phase' models where mostly a method is applied (called 'upwind or donor cell differencing scheme') with the mean parameter values to be shifted (in flow direction) to the node boundaries in CCM a more detailed mixture-fluid approach is asked. This is also demanded because, as to be seen later-on from the relations of the sections 3.7 to 3.9, not only absolute nodal SC boundary or mean nodal function values are required but as well also their nodal slopes f^{s}_{Nn} and f^{s}_{Mn} together with their gradients f^{z}_{Nn} since according to this approach the length of SC nodes can tend also to zero.

$$f^{s}_{Nn} = \frac{(f_{Nn} - f_{Nn-1})}{\Delta z_{Nn}} \rightarrow f^{z}_{CEI}\left(\text{at } n=1\right) = \text{input} \ \text{ or } \rightarrow f^{s}_{Nn-1}\left(\text{at } n=N_{CT}>1\right) \text{ if } Dz_{Nn} \rightarrow 0 \tag{16}$$

$$f^{s}_{Mn}=2\frac{(f_{Mn}-f_{Nn-1})}{\Delta z_{Nn}} \rightarrow f^{z}_{CEI}(\text{at } n=1)=\text{input or} \rightarrow f^{z}_{Nn-1}\ (\text{at } n=N_{CT}>1) \text{ if } \Delta z_{Nn} \rightarrow 0 \qquad (17)$$

Hence, for this purpose a special 'quadratic polygon approximation' procedure, named 'PAX', had to be developed. It plays (together with the Leibniz rule presented above) an outstanding part in the development of the here presented 'mixture-fluid model' and helps, in particular, to solve the difficult task of how to take care of varying SC boundaries (which can eventually cross BC node boundaries) in an appropriate and exact way.

3.3.1. Establishment of an effective and adequate approximation function

The PAX procedure is based on the assumption that the solution function f(z) of a PDE (for example temperature or void fraction) is split into a number of N_{CT} nodal SC functions $f_n(z,t)$. Each of them has then to be approximated by a specially constructed quadratic polygon which have to fulfil the following requirements:

- The node entrance functions (f_{Nn-1}) must be either equal to the SC entrance function ($f_{Nn-1} = f_{CE}$) (if n = 1) or to the outlet function of the node before (if n > 1). This is obviously not demanded for gradients of the nodal entrance functions (except for the last node at $n = N_{CT}$).
- The mean function values f_{Mn} over all SC nodes have to be preserved (otherwise the balance equations could be hurt).
- With the objective to guarantee stable behaviour of the approximated functions (for example by excluding 'saw tooth-like behaviour') it will, in an additional assumption, be demanded that the outlet gradients of the first N_{CT} -1 nodes should be set equal to the slopes between their neighbour mean function values. The entrance gradient of the last node ($n= N_{CT}$) should be either equal to the outlet gradient of the node before (if $n = N_{CT} > 1$) or equal to a given SC input gradient (for the special case $n = N_{CT} =1$). Thus

$$\begin{aligned} f^{z}_{Nn}=(\frac{\partial f}{\partial z})_{Nn} &= 2\frac{f_{Mn+1}-f_{Mn}}{\Delta z_{Nn+1}+\Delta z_{Nn}} \qquad (n=1, N_{CT}-1, \text{ if } N_{CT}>1) \\ &= \frac{2}{\Delta z_{Nn}}\left(2f_{Nn} - 3f_{Mn} + f_{Nn-1}\right) \rightarrow f^{(z)}_{Nn-1} \text{if} \Delta z_{CA} 0 \ (n = N_{CT}, \text{ if } N_{CT} > 1) \end{aligned} \qquad (18)$$

$$\begin{aligned} f^{z}_{Nn-1}=f^{z}_{CE}=f^{z}_{CEI}=\frac{2}{\Delta z_{Nn}}(3f_{Mn}-f_{CA} - 2f_{CE}) \qquad & \left(n = N_{CT}=1\right) \\ =^{z}_{Nn}\left(\text{of the node before}\right) \qquad & \left(n = N_{CT}> 1\right) \end{aligned} \qquad (19)$$

This means, the corresponding approximation function reaches not only over the node n. Its next higher one (n+1) has to be included into the considerations too (except, of course, for the last node). This assumption makes the PAX procedure very effective (and stable). It is a conclusive onset in this method since it helps to smooth the curve, guarantees that the gradients at the upper or lower SC boundary do not show abrupt changes if these boundaries cross a BC node boundary and has the effect that perturbations at channel entrance do not directly affect corresponding parameters of the upper BC nodes.

For the special case of a SC having shrunk to a single node ($n=N_{CT}=1$) the quadratic approximation demands as an additional input to PAX (instead of the now not available term f_{Mn}) the gradient f^{z}_{CEI} at SC entrance. It represents thereby the gradient of either the coolant temperature T^{z}_{CEI} or void fraction α^{z}_{CEI} (in case of single- or two-phase flow entrance conditions). If this parameter is not directly available it can, for example, be estimated by combining the mass and energy balance equations at SC entrance in an adequate way (See Hoeld, 2005). This procedure allows to take care not only of SC-s consisting of only one single node but also of situations where during a transient either the first or last SC of a BC starts to disappear or to be created anew (i.e. $z_{CA} \rightarrow z_{BE}$ or $z_{CE} \rightarrow z_{BA}$), since now the nodal mean value f_{Mn} at $n = N_{CT}$ (for both $N_{CT} = 1$ or > 1) is no longer or not yet known.

3.3.2. Resulting nodal parameters due to PAX

It can be expected that in the steady state case after having solved the basic set of non-linear algebraic equations (as presented later-on in the sections 3.7, 3.8 and 3.9) as input to PAX the following parameters will be available:

- Geometry data such as the SC entrance (z_{CE}) and node positions (z_{Nn}) (and thus also the SC outlet boundary position z_{CA} as explained in section 3.9) determining then in PAX the number of SC nodes (N_{CT}),
- the SC entrance function $f_{N0} = f_{CE}$ and (at least for the special case $n=N_{CT}=1$) its gradient $f^{(z)}_{CEI}$

and finally

- the nodal boundary functions f_{Nn} ($n=1,N_{CT}$) with $f_{CA} = f_{Nn}$ at $n = N_{CT}$ and $f_{CA}= f_{LIMCA}$ if $z_{CA} < z_{BA}$.

Based on these inputs PAX yields then the nodal mean function values f_{Mn} (at $n=1,N_{CT}$)

$$
\begin{aligned}
f_{Mn} &= \frac{(\Delta z_{Nn+1}+\Delta z_{Nn})(2f_{Nn}+f_{Nn-1})-\Delta z_{Nn} f_{Mn+1}}{3\Delta z_{Nn+1}+2\Delta z_{Nn}} \\
&\quad (n=1, N_{CT}-1, N_{CT}>1 \text{ if } z_{CA}=z_{BA} \text{ or } n=1, N_{CT}-2, N_{CT}>2 \text{ if } z_{CA}<z_{BA}) \\
&= \frac{1}{3}(f_{CA}+2 f_{CE})+\frac{1}{6} Dz_{CA} f^{(z)}_{CEI} \qquad (n = N_{CT}= 1) \\
&= \frac{1}{3}(f_{CA}+2 f_{CE})+\frac{1}{6}\Delta z_{CA} f(f_{CA}- f_{Nn-2}) \qquad (f_{CA}- f_{Nn-2})
\end{aligned}
\tag{20}
$$

functions which are needed as initial values for the transient case.

In the transient case it can be expected that after having integrated the set of non-linear ordinary differential equations (ODE-s) (as to be shown again in the sections 3.7, 3.8 and 3.9) the SC outlet position z_{CA} ($=z_{Nn}$) and thus also the total number N_{CT} of SC nodes will now be directly available from the integration. Thereby, as input data needed for the use in the PAX procedure it has now to be distinguished between two cases:

- if the now known SC outlet position is identical with the BC outlet ($z_{CA}=z_{BA}$) the mean nodal function values f_{Mn} for all N_{CT} nodes ($n=1,N_{CT}$) should be provided as input

or

- if the SC outlet position moves still within the BC ($z_{CA} < z_{BA}$) the mean nodal function values f_{Mn} of only N_{CT}-1 nodes ($n=1, N_{CT}-1$) are demanded. Additionally, the nodal function limit values f_{LIMNn} (usually saturation temperature values at single-phase flow resp. void fraction values limited by 1 or 0 at mixture flow conditions) are asked. Then the missing mean nodal function f_{Mn} of the last SC is not any longer needed, since this function can be determined from eq.(20) in dependence of the known outlet function $f_{Nn} = f_{CA} = f_{LIMCA}$ at $z_{Nn} = z_{CA}$.

These nodal input function values can then be taken (together with its input parameter f_{CE} and the nodal positions z_{BE} and z_{Bn} at $n=1, N_{CT}$) as basic points for the PAX procedure, yielding, after rearranging the definition equations of the approximation function in an adequate way, all the other not directly known nodal function parameters of the SC.

Hence, it follows for the special situation of a SC being the last one within the BC ($z_{CA} = z_{BA}$)

$$
\begin{aligned}
f_{Nn} &= \frac{1}{2}\left(3f_{Mn} - f_{Nn-1}\right) + \frac{1}{2}\frac{\Delta z_{Nn}}{\Delta z_{Nn+1}+\Delta z_{Nn}}\left(f_{Mn+1} - f_{Mn}\right) && (n=1,\ N_{CT}-1 \text{ with } N_{CT}>1 \text{ if } z_{CA}=z_{BA}) \\
&= 3f_{Mn} - 2f_{CE} - \frac{1}{2}\Delta z_{CA} f^{z}_{CEI} && (n = N_{CT} = 1 \text{ if } z_{CA}=z_{BA}) \\
&= f_{CA} = 2\left(f_{Mn} - f_{Mn-1}\right) + f_{Nn-2} && (n = N_{CT} > 1 \text{ if } z_{CA}=z_{BA})
\end{aligned}
\qquad (21)
$$

resp. for the case $z_{CA} < z_{BA}$

$$
\begin{aligned}
f_{Nn} &= \frac{1}{2}\left(3f_{Mn} - f_{Nn-1}\right) + \frac{1}{2}\frac{\Delta z_{Nn}}{\Delta z_{Nn+1}+\Delta z_{Nn}}\left(f_{Mn+1} - f_{Mn}\right) && (n=1,\ N_{CT}-2 \text{ with } N_{CT}>2 \text{ if } z_{CA}<z_{BA}) \\
&= \frac{1}{2}\left(3f_{Mn} - f_{Nn-1}\right) + \frac{1}{4}\frac{\Delta z_{Nn}}{\Delta z_{Nn+1}+\Delta z_{Nn}}\left(f_{LIMCA} - f_{Nn-1}\right) && (n=N_{CT}-1 \text{ with } N_{CT}>1 \text{ if } z_{CA}<z_{BA}) \\
&= f_{CA} = f_{LIMCA} && (n = N_{CT} \quad \text{if } z_{CA}<z_{BA})
\end{aligned}
\qquad (22)
$$

The last mean nodal function f_{Mn} (at $n=N_{CT}$) which is, for this case ($z_{CA} < z_{BA}$), not available from the integration, follows now directly from eq.(21) if replacing there f_{CA} by f_{LIMCA}.

Finally, from the eqs.(16), (17) and (19) the slopes and gradients can be determined.

The corresponding time-derivative of the last mean node function which is needed for the determination of the SC boundary time-derivative (see section 3.9) follows (as long as $z_{CA} < z_{BA}$) by differentiating the relation above

$$\frac{d}{dt}f_{Mn} = f^{t}_{PAXCA} + f^{z}_{PAXCA}\frac{d}{dt}z_{CA} \quad \left(n = N_{CT} \text{ if } z_{CA} < z_{BA}\right) \tag{23}$$

with the coefficients

$$\begin{aligned} f^{t}_{PAXCA} &= \frac{1}{3}\left(\frac{d}{dt}f_{LIMCA} + 2\frac{d}{dt}f_{CE}\right) - \frac{1}{6}\left(f^{z}_{CEI}\frac{d}{dt}z_{CE} - Dz_{CA}\frac{d}{dt}f^{z}_{CEI}\right) && (n = N_{CT} = 1 \text{ if } z_{CA} < z_{BA}) \\ &= \frac{d}{dt}f_{Mn-1} + \frac{1}{2}\frac{d}{dt}f_{LIMCA} - \frac{1}{2}\frac{d}{dt}f_{Nn-2} && (n = N_{CT} > 1 \text{ if } z_{CA} < z_{BA}) \\ f^{z}_{PAXCA} &= \frac{1}{6}f^{(z)}_{CEI} \text{ or } = 0 && \left(n = N_{CT} = 1 \text{ or } > 1 \text{ if } z_{CA} < z_{BA}\right) \end{aligned} \tag{24}$$

The differentials $\frac{d}{dt}f_{Mn-1}$, $\frac{d}{dt}z_{CA}$, $\frac{d}{dt}f_{LIMCA}$ are directly available from CCM and, if N_{CT}=2, the term $\frac{d}{dt}f_{Nn-2} = \frac{d}{dt}f_{CE}$ from input too. For the case that a SC contains more than two nodes only their corresponding mean values are known, the needed term $\frac{d}{dt}f_{Nn-2}$ has thus to be estimated by establishing the time-derivatives of all the boundary functions at the nodes below $N_{CT} < 2$.

3.3.3. Code package PAX

Based on the above established set of equations a routine PAX had to be developed. Its objective was to calculate automatically either the nodal mean or nodal boundary values (in case of an either steady state or transient situation). The procedure should allow also determining the gradients and slopes at SC entrance and outlet (and thus also outlet values characterizing the entrance parameters of an eventually subsequent SC). Additionally, contributions needed for the calculation of the time-derivatives of the boiling boundary or mixture level can be gained (See later-on the eqs.(66) and (67)).

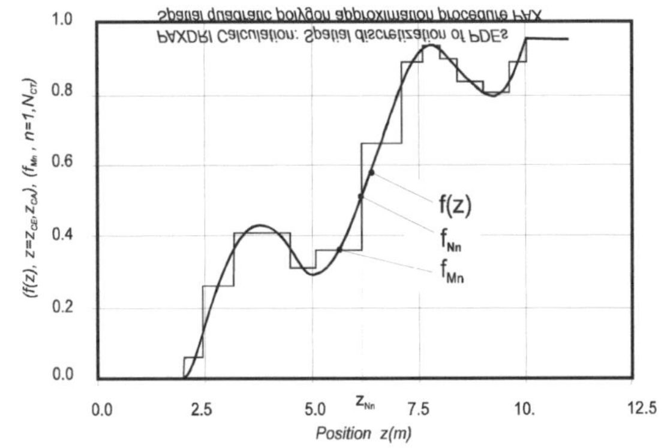

Figure 2. Approximation function f(z) along a SC for both steady state and transient conditions after applying PAX (Example)

Before incorporating the subroutine into the overall coolant channel module the validity of the presented PAX procedure has been thoroughly tested. By means of a special driver code (PAXDRI) different characteristic and extreme cases have been calculated. The resulting curves of such a characteristic example are plotted in fig.2. It represents the two approximation curves of an artificially constructed void fraction distribution $f(z) = \alpha(z)$ along a SC with two-phase flow both for the steady state but also transient situation. Both curves (on the basis of f_{Mn} and f_{Nn}) should be (and are) identical.

3.4. Needed input parameters

3.4.1. Initial conditions

For the start of the transient calculations adequate steady state parameters have to be available as initial conditions.

3.4.2. Boundary conditions

As boundary conditions for both the steady state and especially for transient calculations the following input parameters are expected to be known. Thereby demanding only easily available BC values (They will, within CCM, then be automatically translated into the corresponding SC values):

- Power profile along the entire BC. This means that either the nodal heat flux terms q_{FBE} and q_{FBk} (at BC entrance and each node $k=1,N_{BT}$) or q_{FBE} and the nodal power terms Q_{BMk} are expected to be known, either directly from input or (as explained in section 2.2.4) by solving the appropriate 'Fourier heat conduction equation'. From the relation

$$Q_{BMk}=\frac{1}{2}\Delta z_{Bk}\left(q_{LBk}+q_{LBk-1}\right)=\frac{1}{2}\Delta z_{Bk}\left(U_{TWBk}q_{FBk}+U_{TWBk-1}q_{FBk-1}\right)=V_{BMk}q_{BMk}$$

$$\text{with } q_{LBE}=U_{TWBE}q_{FBE}\text{ and } q_{DBE}=\frac{U_{TWBE}}{A_{BE}}q_{FBE} \qquad (k=1,N_{BT}) \qquad (25)$$

then the other BC nodal terms (q_{FBk} or Q_{BMk}, q_{LBk} and q_{DBk}) can be determined too. (Hoeld, 2002b, 2004a and 2011).

- For normalization purposes at the starting calculation (i.e., at the steady state situation) as an additional parameter the total nominal (steady state) heat power $Q_{NOM,0}$ is asked.
- Channel entrance temperature T_{BEIN} (or enthalpy h_{BEIN})
- System pressure P_{SYS} and its time-derivative (dP_{SYS}/dt), situated at a fixed position either along the BC (entrance, outlet) or even outside of the ensemble. Due to the fast pressure wave propagation each local pressure time-derivative can then be set equal to the change in system pressure (as described in section 3.6).
- Total mass flow G_{BEIN} at BC entrance together with pressure terms at BC entrance P_{BEIN} and outlet P_{BAIN}. These three parameters are needed for steady state considerations (and partially

used for normalization purposes). In the transient case only two of them are demanded as input. The third one will be determined automatically by the model. These allows then to distinguish between the situation of an 'open' or 'closed channel' concept as this will be explained in more detail in section 3.11.

- Steam mass flow G_{SBEIN} at BC entrance (=0 or = G_{BEIN} at single- or $0 < G_{SBEIN} < G_{BEIN}$ at two-phase flow conditions). The corresponding entrance void fraction α_{BE} will then be determined automatically within the code by applying the inverse drift-flux correlation.

Eventually needed time-derivatives of such entrance functions can either be expected to be known directly from input or be estimated from their absolute values.

By choosing adequate boundary conditions then also thermal-hydraulic conditions of other situations can be simulated. For example that of several channel assembles (nuclear power plants, test loops etc.) which can consist of a complex web of pipes and branches (represented by different BC-s, all of them distinguished by their key numbers KEYBC). Even the case of an ensemble consisting of inner loops (for example describing parallel channels) can be treated in an adequate way according to the concept of a 'closed' channel (see section 3.11).

3.4.3. *Solution vector*

The characteristic steady state parameters are determined in a direct way, i.e. calculated by solving the non-linear set of algebraic equations for SC-s (as being presented in the chapters 3.7, 3.8 and 3.9). Thereby, due to the nonlinearities in the set of the (steady state) constitutive equations a recursive procedure in combination with and controlled by the main program has to be applied until a certain convergence in the solution vector can be stated. The results are then combined to BC parameters and transferred again back to the main (= calling) program.

For the transient case, as a result of the integration (performed within the calling program and thus outside of CCM) the solution parameters of the set of ODE-s are transferred after each intermediate time step to CCM. These are (as described in detail also in chapter 4) mainly the mean nodal SC and thus BC coolant temperatures, mean nodal void fractions and the resulting boiling or superheating boundaries. These last two parameters allow then to subdivide the BC into SC-s yielding the corresponding constitutive parameters and the total and nodal length (z_{Nn} and Δz_{Nn}) of these SC-s and thus also their total number (NCT) of SC nodes. Finally, the needed SC (and thus BC) time-derivatives can be determined within CCM (as described in the sections 3.7, 3.8 and 3.9) and then transmitted again to the calling program where the integration for the next time step can take place.

3.5. SC power profile

Knowing (as explained in section 3.4.2) the nodal BC power Q_{BMk} together with the linear power q_{LBE} at BC entrance the corresponding SC power profile can now be determined too.

Hence, since linear behaviour of the linear nodal power terms within the corresponding BC nodes can be assumed it follows for the 'linear SC power' term

$$\begin{aligned} q_{LNn} = q_{LCE} &= q_{LBE}\left(=\text{BC entr.}\right) \text{ or } = \left(q_{LCA}\right)_{\text{of the last node of the SC before}} \left(n=0 \text{ if } z_{CE} = \text{or} > z_{BE}\right) \\ &= q_{LBk} \qquad \left(n=1,N_{CT} \text{and } k=n+N_{BCE} \text{if } L_{FTYPE}=2\right) \\ &= q_{LBk} \qquad \left(n=1,N_{CT}-1 \text{and, if } z_{CA}=z_{BA},\ n=N_{CT} \text{with } k=n+N_{BCE}\right) \\ &= q_{LCA} = q_{LBk-1} + \left(q_{LBk} - q_{LBk-1}\right)\frac{\Delta z_{CA}}{\Delta z_{Bk}} \left(n=N_{CT} \text{and } k=N_{BCA} \text{if } z_{CA} < z_{BA}\right) \end{aligned} \tag{26}$$

for the 'nodal SC power term'

$$\begin{aligned} Q_{Mn} &= \frac{1}{2}\Delta z_{Nn}\left(q_{LNn} + q_{LNn-1}\right) && \left(n=1,N_{CT}\right) \\ &= Q_{BMk} - \left(Q_{MCA}\right)_{\text{of the last node of the SC before}} && \left(n=1 \text{ with } k=1+N_{BCE}\right) \\ &= Q_{BMk} && \left(n=2,\ N_{CT}-1 \text{and } k=n+N_{BCE}\right) \\ &= Q_{MCA} = Dz_{CA}\left[q_{LBk-1} + \frac{1}{2}\left(q_{LBk} - q_{LBk-1}\right)\left(q_{LBk} - q_{LBk-1}\right)\frac{\Delta z_{CA}}{\Delta z_{Bk}}\right] && \left(n=N_{CT} \text{and } k=N_{BCA} \text{if } z_{CA} < z_{BA}\right) \end{aligned} \tag{27}$$

and the 'mean nodal' and 'nodal boundary SC power density' terms (q_{Mn} and q_{Nn})

$$\begin{aligned} q_{Mn} &= \frac{Q_{Mn}}{V_{Mn}} \text{ and (independently of the node length)} = \frac{1}{2A_{Mn}}\left(q_{LNn-1} + q_{LNn}\right) \ \left(n=1,N_{CT}\right) \\ q_{Nn} &= q_{CE} = q_{BE} \text{ or } = \left(q_{CA}\right)_{\text{of the last node of the SC before}} \qquad \left(n=0 \text{ if } z_{CE} = \text{or} > z_{BE}\right) \\ &= 2q_{Mn} - q_{Nn-1} \qquad \left(n=1,N_{CT}\right) \end{aligned} \tag{28}$$

These last parameters are, since independent of Δz_{Nn}, very useful for equations where it has to be expected that Δz_{Nn} -> 0 (as this can be seen later-on by the eqs.(32) and (49)).

The SC length z_{CA} (and thus that of its last node Δz_{CA}) are (for the case $z_{CA,0} < z_{BA}$) only known in the transient case (as a result of the integration procedure). For steady state conditions the term $Q_{MCA,0}$ follows from energy balance considerations (eqs.(40) and (59)). Then, in a reverse manner, $\Delta z_{CA,,0}$ can be calculated (See eq.(65) in section 3.9).

3.6. Decoupling of mass and energy balance from momentum balance equations

Treating the conservation equations in a direct way produces due to elements with fast pressure wave propagation (which are responsible for very small time constants) a set of 'stiff' ODE-s. This has the consequence that their solution turns out to be enormously CPU-time consuming. Hence, to avoid this costly procedure CCM has been developed with the aim to decouple the mass and energy from their momentum balance equations. This can be achieved by determining the thermodynamic properties of water and steam in the energy and mass balance equations on the basis of an estimated pressure profile P(z,t). Thereby the pressure difference terms from a recursive (or a prior computational time step) will be added to an eventually time-varying system pressure $P_{SYS}(t)$, known from boundary conditions. After having solved the two conservation equations for mass and energy (now separately from and not simultaneously with the momentum balance) the different nodal pressure gradient terms can (by the then following momentum balance considerations) be determined according to the eqs.(4), (5) and (6).

It can additionally be assumed that according to the very fast (acoustical) pressure wave propagation along a coolant channel all the local pressure time-derivatives can be replaced by a given external system pressure time-derivative, i.e.,

$$\frac{d}{dt}P(z,t) \cong \frac{d}{dt}P_{SYS} \tag{29}$$

By applying the above explained 'intelligent' (since physically justified) simplification in CCM the small, practically negligible, error in establishing the thermodynamic properties on the basis of such an estimated pressure profile can be outweighed by the enormous benefit substantiated by two facts:

- Avoidance of the very time-consuming solution of stiff equations,
- the calculation of the mass flow distribution into different channels resulting from pressure balance considerations can, in a recursive way, be adapted already within each integration time step, i.e. there is no need to solve the entire set of differential equations for this purpose (See 'closed channel' concept in section 3.11).

3.7. Thermal-hydraulics of a SC with single-phase flow ($L_{FTYPE} > 0$)

The spatial integration of the two PDE-s of the conservation eqs.(1) and (2) over a (single-phase) SC node n yields (by taking into account the rules from section 3.2, the relations from the eqs. (7) and (29), the possibility of a locally changing nodal cross flow area along the BC and the fact that eventually V_{Mn} -> 0) for the transient case the relation

- a relation for the total nodal mass flow

$$\begin{aligned} G_{Nn} = G_{Nn-1} - V_{Mn}(\varrho^{T}_{Mn}\frac{d}{dt}T_{Mn} + \varrho^{P}_{Mn}\frac{d}{dt}P_{SYS}) + (r_{Nn}-r_{Mn})A_{Nn}P_{SYS}) + (r_{Nn}-r_{Mn})A_{Nn}\frac{d}{dt}z_{Nn} \\ + (r_{Mn}-r_{Nn-1})A_{Nn-1}\frac{d}{dt}z_{Nn-1} \qquad (n=1,N_{CT}),\ L_{FTYPE}>0 \end{aligned} \tag{30}$$

and

- the time-derivative for the mean nodal coolant temperature (if eliminating the term G_{Nn} in the resulting equation by inserting from the equation above):

$$\begin{aligned} \frac{d}{dt}T_{Mn} = T^{t}_{Tn} + T^{z}_{TCA}\frac{d}{dt}z_{Nn} \text{ with } \frac{d}{dt}z_{Nn} = \frac{d}{dt}z_{CA} \text{ or } = 0 \text{ if } n = N_{CT} \text{ or } < N_{CT} \\ (n=1, N_{CT} \text{ and } z_{CA}=z_{BA}) \text{ or } (n=1, N_{CT}-1 \text{ and } z_{CA}<z_{BA}),\ L_{FTYPE}>0 \end{aligned} \tag{31}$$

with the coefficients

$$T^{t}_{Tn}=\frac{q_{Mn}-q_{Gn}+q_{Pn}}{\varrho_{Mn}h^{T}_{Mn}C_{TMn}}+T\frac{d}{dt}z_{Nn-1}\ \text{with}\frac{d}{dt}z_{Nn-1}=\frac{d}{dt}z_{CE}\ \text{or}=0\ \text{if}\ n=1\ \text{or}>1$$
$$=\frac{1}{V_{Mn}\varrho_{Mn}h^{T}_{Mn}C_{TMn}}[Q_{Mn}-G_{Nn-1}(h_{Nn}-h_{Nn-1})+V_{Mn}q_{Pn}]+T+T^{z}_{TCE}\frac{d}{dt}z_{Nn-1} \tag{32}$$

$$C_{TMn}=1-\frac{\rho^{T}_{Mn}}{\rho_{Mn}}\frac{h_{Nn}-h_{Mn}}{h^{T}_{Mn}}=1-\frac{\rho^{T}_{Mn}}{\rho_{Mn}}\left(T_{Nn}-T_{Mn}\right) \tag{33}$$

$$q_{Gn}=\frac{G_{Nn-1}}{A_{Mn}}h^{s}_{Nn} \tag{34}$$

$$q_{Pn}=C_{RHPn}\frac{d}{dt}P_{SYS}\ \text{and}\ C_{RHPn}=1-r_{Mn}h^{P}_{Mn}+\rho^{P}_{Mn}\left(h_{Nn}-h_{Mn}\right) \tag{35}$$

$$q_{Zn}=\frac{1}{2A_{Mn}}\{A_{Nn}r_{Mn}h^{(s)}_{Nn}\frac{d}{dt}z_{Nn}+A_{Nn-1}[\rho_{Mn}h^{(s)}_{Mn}-2(\rho_{Mn}-\rho_{Nn-1})\,h^{(s)}_{Nn})]\frac{d}{dt}z_{Nn-1}\} \tag{36}$$

$$T^{z}_{TCE}=\frac{1}{2}[T^{s}_{Mn}-2(1-\frac{\rho_{CE}}{\rho_{Mn}})T^{s}_{Nn}]\frac{A_{CE}}{A_{Mn}C_{TMn}}\ \text{or}\ =0\ \text{at}\ n=1\ \text{and}\ L_{FTYPE}>0$$
$$(\text{if either}\ z_{CE}>z_{BE}\ \text{or}\ z_{CE}=z_{BE}) \tag{37}$$

$$T^{z}_{TCA}=0\ \text{or}\ =\frac{1}{2}\frac{A_{Nn}}{A_{Mn}C_{TMn}}T^{s}_{Nn}$$
$$\left(n<N_{CT}\ \text{and}\ z_{CA}<z_{BA}\ \text{or}\ n=N_{CT}\ \text{and}\ z_{CA}=z_{BA}\right),\ L_{FTYPE}>0 \tag{38}$$

In the transient case the mean nodal coolant temperature value T_{Mn} is at the begin of each (intermediate) time step known. This either, at the first time step, from steady state considerations (in combination with PAX) or as a result of the integration procedure. Hence, additional parameters needed in the relations above can be determined too. From the PAX procedure it follow also the SC nodal terms T_{Nn}, $T^{(s)}_{Nn}$ and the slopes $T^{(s)}_{Mn}$ resp., for the case that $\Delta z_{Nn} \rightarrow 0$, their gradients. Finally, using the water/steam tables (Hoeld, 1996), also their nodal enthalpies are fixed.

From the integration procedure then also the SC outlet position z_{CA} is provided, allowing to determine the total number of SC nodes (N_{CT}) too. The situation that $z_{CA} = z_{BT}$ means the SC nodal boundary temperature values have, within the entire BC, not yet reached their limit values (T_{LIMNn}=T_{SATNn}). Thus N_{CT}= N_{BCA} with N_{BCA}=N_{BT}–N_{BCE}. Otherwise, if $z_{CA} < z_{BT}$ this limit is reached (at node n), then N_{CT} = n. Obviously, the procedure above yields also the time-derivative of the SC outlet position moving within this channel (As described in section 3.9).

The steady state part of the total nodal mass flow (charaterized by the index 0) follows from the basic non-linear algebraic equation (30) if setting there the time-derivative equal to 0:

$$G_{Nn,0} = G_{CA,0} = G_{CE,0} = G_{BA,0} = G_{BE,0} \qquad \left(n=1,N_{CT}\right),\ L_{FTYPE}>0 \tag{39}$$

Treating eq.(31) in a similar way and multiplying the resulting relation by $V_{Mn,0}$ yields the steady state nodal temperature resp. enthalpy terms

$$h_{Nn,0} = h_{Nn-1,0} + \frac{Q_{Mn,0}}{G_{BE,0}} \leq h'_{Nn,0} \text{ or} \geq h''_{Nn,0} \qquad \left(\text{with } h_{Nn-1,0} = h_{CE,0} \text{at } n=1\right)$$
$$\left(\text{if } L_{FTYPE} = 1 \text{ or} = 2 \text{ at } n = 1,\ N_{BT}\text{-}N_{BCE}\right) \tag{40}$$

restricted by their saturation values. Then, with regard to eq.(27) which results from energy balance considerations, the nodal power term for the last SC node has to obey the relation

$$Q_{MCA,,0} = Q_{Mn,,0} = (h'_{CA,0} - h_{Nn-1,0})\, G_{BE,0} \qquad \left(\text{if } n = N_{CT} < N_{BCA} \text{at } L_{FTYPE} = 1\right)$$
$$= (h_{Nn-1,0} - h''_{CA,0})\, G_{BE,0} \qquad \left(\text{if } n = N_{CT} < N_{BCA} \text{at } L_{FTYPE} = 2\right) \tag{41}$$

Thus, for the steady state case, N_{CT} is fixed too:

$$N_{CT} = n < N_{BCA} = N_{BT} - N_{BCE} + 1 \text{ and } z_{CA,0}\left(=z_{Nn,0}\right) < z_{BA} \ \ (\text{if } h_{Nn,0} = h'_{Nn,0} \text{ at } L_{FTYPE} = 1)$$
$$N_{CT} = n = N_{BCA} \text{ and } z_{CA,0} = z_{BA} \qquad (\text{if } h_{Nn,0} < h'_{Nn,0} \text{ at } L_{FTYPE} = 1) \tag{42}$$

with similar relations for the case L_{FTYPE} = 2.

From the resulting steady state enthalpy value $h_{Nn,0}$ follows then (by using the thermodynamic water/steam tables) the corresponding coolant temperature value $T_{Nn,0}$ (with $T_{Nn,0} = T_{SATNn,0}$ if $n = N_{CT}$ and $z_{CA} < z_{BA}$) and, by applying the PAX procedure (see section 3.3), their mean nodal temperature and enthalpy values $T_{Mn,0}$ and $h_{Mn,0}$, parameters which are needed as start values for the transient calculations. Obviously, due to the non-linearity of the basic steady state equations, this procedure has to be done in a recursive way.

It can additionally be stated that both the steady state and transient two-phase mass flow parameters get the trivial form

$$G_{SNn} = G_{SNn,0} = 0 \text{ resp. } G_{WNn} = G_{CE} \text{ and } G_{WNn,0} = G_{BE,0} \quad \left(n=1,\ N_{CT},\ \text{if } L_{FTYPE} = 1\right)$$
$$G_{SNn} = G_{CE} \text{ and } G_{SNn,0} = G_{BE,0} \text{ resp. } G_{WNn} = G_{WNn,0} = 0 \quad \left(n=1,\ N_{CT},\ \text{if } L_{FTYPE} = 2\right) \tag{43}$$

And

$$\frac{d}{dt}\alpha_{Mn} = 0\,,\ \alpha_{Nn} = \alpha_{Nn,0} = \alpha_{Mn,0} = 0 \quad \text{or} = 1 \text{ and } X_{Nn,0} = \frac{h_{Nn,0} - h'_{Nn,0}}{h_{SWNn,0}} \quad \left(n=1,\ N_{CT},\ \text{if } L_{FTYPE}=1 \text{ or } = 2\right) \tag{44}$$

3.8. Thermal-hydraulics of a SC with two-phase flow ($L_{FTYPE} = 0$)

The spatial integration of the two PDE-s of the conservation) eqs.(1) and (2) (now over the mixture-phase SC nodes n) can be performed by again taking into account the rules from section 3.2, the relations from the eqs.(7) and (29), by considering the possibility of locally changing nodal cross flow areas along the BC) and the fact that eventually V_{Mn} -> 0. This yields then relations for

- the total mass flow term

$$G_{Nn} = G_{Nn-1} + V_{Mn}(\rho' - \rho'')_{Mn}\left(\frac{d}{dt}\alpha_{Mn} - \alpha^{t}_{GPn} - \alpha^{t}_{GZn}\right) \quad \left(n=1, N_{CT},\ L_{FTYPE} = 0\right) \tag{45}$$

with, if neglecting thereby small differences between mean and nodal thermodynamic saturation values, the coefficients

$$\alpha^{t}_{GPn} = \frac{1}{(\varrho' - \varrho'')_{Mn}}\left[(1-\alpha)\rho'^{/P} + \alpha\rho''^{/P}\right]_{Mn}\frac{d}{dt}P_{SYS} \tag{46}$$

$$\begin{aligned} \alpha^{t}_{GZn} &= \alpha^{z}_{CE}\frac{d}{dt}z_{CE} = \frac{1}{2}\frac{A_{CE}}{A_{Mn}}\alpha^{s}_{Mn}\frac{d}{dt}z_{CE} && (n = 1 \text{ and } z_{CE} > z_{BE}) \\ &= 0 && (1 < n < N_{CT}) \\ &= \alpha^{z}_{CA}\frac{d}{dt}z_{CA} = \frac{A_{CA}}{A_{Mn}}\left(\alpha^{s}_{CA} - \frac{1}{2}\alpha^{s}_{Mn}\right)\frac{d}{dt}z_{CA} && (n = N_{CT},\ N_{CT} > 1 \text{ and } z_{CA} < z_{BA}) \end{aligned} \tag{47}$$

and

- the mean nodal void fraction time-derivative

$$\frac{d}{dt}\alpha_{Mn} = \alpha^{t}_{ASn}\alpha^{t}_{APn} + \alpha^{z}_{CA}\frac{d}{dt}z_{Nn} + \alpha^{z}_{CE}\frac{d}{dt}z_{Nn-1} \quad \left(n=1, N_{CT} \text{ and } z_{CA} = z_{BA}\right) \text{ or } \left(n=1,\ N_{CT}-1,\ N_{CT}>1 \text{ and } z_{CA} < z_{BA}\right),\ L_{FTYPE}=0 \tag{48}$$

with the coefficients

$$\alpha^{t}_{ASn}=\alpha^{t}_{AQn}-\alpha^{t}_{AGn}=\frac{1}{\varrho^{//}_{Mn}\,h_{SWMn}}\left[q_{Mn}-h_{SWMn}\frac{G^{s}_{SNn}}{A_{Mn}}\right]$$
$$=\frac{1}{V_{Mn}\varrho^{//}_{Mn}\,h_{SWMn}}\left[Q_{Mn}-\left(G_{SNn}-G_{SNn-1}\right)h_{SWMn}\right] \qquad (49)$$

$$\alpha^{t}_{APn}=C_{RHPn}\frac{d}{dt}P_{SYS}\text{ with }C_{RHPn}=\frac{1}{(\rho^{//}h_{SW})_{Mn}}\left[(1-a)\rho\acute{h}^{\prime P}+\alpha(\rho\acute{h}^{\prime\prime P}+\rho^{\prime\prime P}h_{SW})-1\right]_{Mn} \qquad (50)$$

$$\alpha^{t}_{AZn}=\alpha^{t}_{GZn}\quad\left(\text{see eq.}(47)\right) \qquad (51)$$

$$G^{s}_{SNn}=\frac{\Delta G_{SNn}}{\Delta z_{Nn}}\rightarrow G^{z}_{SNn}=G^{\alpha}_{SNn}\alpha^{s}_{Nn}\text{ if }\alpha_{Nn}\rightarrow\alpha_{CE}=0\text{ (at n=1)}$$
$$\text{or }\alpha_{Nn}\rightarrow\alpha_{CA}=1\text{ (at n}=N_{CT}) \qquad (52)$$

It can again be expected that at the begin of each (intermediate) time step the mean nodal void fraction values α_{Mn} are known. This either from steady state considerations (at the start of the transient calculations) or as a result of the integration procedure. Hence, the additional parameters needed in the relations above can be determined too. From the PAX procedure it follow their nodal boundary void fraction terms α_{Nn} together with their slopes $\alpha^{(s)}_{Nn}$ and $\alpha^{(s)}_{Mn}$ resp. gradient $\alpha^{(z)}_{Nn}$ and thus, as shown both in section 2.2.3 but also in the tables given by Hoeld (2001 and 2002a), all the other characteristic two-phase parameters (steam, water or relative velocities, steam qualities etc). Obviously, due to the non-linearity of the basic equations the steady state solution procedure has to be performed in a recursive way.

If, in the transient case, the SC nodal boundary void fraction α_{Nn} does (within the entire BC) not reach its limit value (α_{LIMNn}=1 or 0) it follows as total number of SC nodes $N_{CT}=N_{BT}-N_{BCE}$ and z_{Nn} (at n=N_{CT}) =z_{CA}=z_{BT}. Otherwise, if this limit is reached (at node n), N_{CT} = n, α_{Nn}=1 (or =0) with z_{CA} (< z_{BT}) resulting from the integration. Then, from the procedure above also the time-derivative of the boiling boundary, moving within BC, can be established (as this will be discussed in section 3.9).

Hence, it follows a relation for the steam mass flow gradients

$$G^{(\alpha)}_{SNn}=A_{CE}v_{S0}\rho^{//}_{Nn}\text{ with }v_{S0}=v_{S}\text{ (at }\alpha_{Nn}=0)\quad(n=1\text{ and }\alpha_{Nn}\rightarrow\alpha_{CE}=0)$$
$$==A_{CA}v_{W1}\rho^{/}_{Nn}\text{ with }v_{W1}=v_{W}(\text{at }\alpha_{Nn}=1)\quad(n=N_{CT}\text{ and }\alpha_{Nn}\rightarrow\alpha_{CA}=1) \qquad (53)$$

If eliminating the term $\frac{d}{dt}\alpha_{Mn}$ in eq.(45) by inserting from eq.(48) yields a relation between G_{SNn} and G_{Nn}

http://dx.doi.org/10.5772/53372

$$G_{Nn} + \left(\frac{\rho^{/}}{\rho^{//}} - 1\right)_{Mn} G_{SNn} = G_{Xn} \quad \left(n=1,N_{CT}, L_{FTYPE}=0\right) \tag{54}$$

The 'auxiliary' mass flow term G_{Xn} is directly available since it refers only to already known parameters (for example taken from the node before or the power profile)

$$G_{Xn} = G_{Nn-1} + \left(\frac{\rho^{/}}{\rho^{//}} - 1\right)_{Mn} \left(G_{SNn-1} + \frac{Q_{Mn}}{h_{SWMn}}\right)) - V_{Mn} \left(\frac{\rho^{//}}{\rho^{/}-}\right)_{Mn} \left(\alpha^{t}_{APn} + \alpha^{t}_{GPn}\right) \tag{55}$$
$$\left(n=1,N_{CT}, L_{FTYPE}=0\right)$$

A similar relation can be established if starting from the drift flux correlation (eq.(10)) and taking advantage of the fact that the needed drift velocity v_{DNn} and the phase distribution parameter C_{0Nn} are independent from the total mass flow G_{Nn} (and can thus be determined before knowing G_{Nn}). This term (G_{Nn}) results then by combining the eqs.(54) and (10)

$$G_{Nn} = \frac{G_{Xn} - (A\,\alpha\, v_D\, \rho^{/}\, C_{DC})_{Nn}}{1 + (\alpha\, C_0\, C_{DC})_{Nn}} \quad (n=1,N_{CT}) \tag{56}$$

with the coefficient

$$C_{DCNn} = \left(\frac{\rho^{/}}{\rho^{//}} - 1\right)_{Mn} \left(\frac{\rho^{//}}{\rho^{/} C_{GC}}\right)_{Nn} \tag{57}$$

From the drift flux correlation package (Hoeld et al., 1992) and (Hoeld, 1994) follow then all the other characteristic two-phase parameters. These are especially the nodal steam mass flow G_{SNn} and, eventually, the slope $\alpha^{(z)}_{Nn}$ resp., according to eq.(53), $G^{(s)}_{SNn}$. Then, finally, from eq. (48) (or eq.(45)) the mean nodal void fraction time-derivative $\frac{d}{dt}\alpha_{Mn}$ will result, needed for the next integration step.

Obviously, at a mixture flow situation the mean nodal temperature and enthalpy terms are equal to their saturation values

$$T_{Mn} = T_{SAT}\left(P_{Mn}\right) \text{ resp. } h_{Mn} = h^{/}\left(P_{Mn}\right) \text{ or } = h^{//}\left(P_{Mn}\right) \quad \left(n=1, N_{CT} \text{and } L_{FTYPE}=0\right) \tag{58}$$

and are thus only dependent on the local resp. system pressure value.

Relations for the steady state case can be derived if setting in the eqs.(45) and (48) (resp. the eq.(49)) the time-derivatives equal to 0. For the total mass flow parameters a similar relation as already given for the single-phase flow (see eq.(48)) is valid, yielding $G_{Nn,0} = G_{BE,0}$.

Hence, one obtains for the (steady state) nodal steam mass flow

$$G_{SNn,0} = G_{SNn-1,0} + \frac{Q_{Mn,0}}{h_{SWMn,0}} \leq G_{Nn,0} = G_{BE,0} \quad \left(n = 1, N_{BT} - N_{BCE} \text{and } L_{FTYPE} = 0\right) \tag{59}$$

if knowing the term $Q_{MCA,,0} = Q_{Mn,,0}$ from eq.(27).

Then the total number (N_{CT}) of SC nodes is given too

$$\begin{array}{lll} N_{CT} = n < N_{BT} - N_{BCE} & \text{and } z_{CA,0}\left(=z_{Nn,0}\right) < z_{BA} & \left(\text{if } G_{SNn,0} = G_{BE,0} \text{and } L_{FTYPE} = 0\right) \\ N_{CT} = N_{BT} - N_{BCE} & \text{and } z_{CA,0} = z_{BA} & (\text{if } G_{SNn,0} < G_{BE,0} \text{and } L_{FTYPE} = 0) \end{array} \tag{60}$$

and also the corresponding steam quality parameter

$$X_{Nn,0} = \frac{G_{SMn,0}}{G_{Nn,0}} \left(n = 1, N_{CT} \text{ if } L_{FTYPE} = 0\right) \tag{61}$$

The nodal boundary void fraction values $\alpha_{Nn,0}$ can now be determined by applying the inverse drift-flux correlation, the mean nodal void fraction value $\alpha_{Mn,0}$ from the PAX procedure. All of them are needed as starting values for the transient calculation.

3.9. SC boundaries

The SC entrance position z_{CE} (= z_{Nn} at n=0) is either (for the first SC within the BC) equal to BC entrance z_{BE} or equal to the SC outlet boundary of the SC before.

In the steady state case the SC outlet boundary (= boiling boundary $z_{BB,0}$ or mixture level $z_{ML,0}$) can be represented as

$$z_{CA,0} = z_{BA} \left(n = N_{CT} \text{and } z_{CA,0} = z_{BA}\right) \tag{62}$$

Or

$$z_{CA,0} = z_{Nn-1} + Dz_{CA,0} \qquad \left(n = N_{CT} \text{and } z_{CA,0} < z_{BA}\right) \tag{63}$$

The (steady state) numbers (N_{CT}) of (single- or two-phase) SC-s are already determined by the eqs.(42) or (60), the nodal power terms $Q_{MCA,0}$ by the eqs.(41) or (59). Hence, after dividing eq. (27) by $\Delta z_{Bk} q_{LBk-1,0}$ one gets an algebraic quadratic equation of the form

$$\frac{1}{2}\frac{q_{LBk,0}-q_{LBk-1,0}}{q_{LBk-1,0}}\left(\frac{\Delta z_{CA,0}}{\Delta z_{Bk}}\right)2+\frac{\Delta z_{CA,0}}{\Delta z_{Bk}}-\frac{Q_{MCA,0}}{z_{Bk}\,q_{LBk-1,0}}=0 \qquad (n=N_{CT} \text{ and } k=N_{BCA} \text{ if } N_{CT}<N_{BCA}) \tag{64}$$

yielding finally as solution

$$\begin{aligned}\Delta z_{CA,0}&=\Delta z_{Bk} \qquad (n=N_{CT} \text{ and } k=N_{BCA} \text{ if } N_{CT}=N_{BT})\\ &=\Delta z_{Bk}\frac{q_{LBk-1,0}}{q_{LBk-1,0}-q_{LBk,0}}\left[1-\sqrt{1-2\left(1-\frac{q_{LBK,0}}{q_{LBk-1,0}}\right)\frac{Q_{CMA,0}}{\Delta z_{Bk}\,q_{LBMk-1,0}}}\right] \qquad (k=N_{BCA} \text{ if } N_{CT}<N_{BT})\\ &\rightarrow\frac{Q_{MCA,0}}{q_{LBk-1,0}}\left[1+\left(1-\frac{q_{LBk,0}}{q_{LBk-1,0}}\right)\frac{Q_{MCA,0}}{z_{Bk}\,q_{LBk-1,0}}\right] \text{ if } q_{LBk,0}\rightarrow q_{LBk-1,0}\end{aligned} \tag{65}$$

Then, from the relations in section 3.5, also the other characteristic steady state power terms can be calculated.

In the transient case the SC outlet boundary z_{CA} (=boiling boundary or mixture level) follows (and thus also Δz_{CA} and N_{CT}), as already pointed-out, directly from the integration procedure. During the transient this boundary can move along the entire BC (and thereby also cross BC node boundaries). A SC can even shrink to a single node (N_{CT} =1), start to disappear or to be created anew. Then, if Δz_{CA} -> 0, in the relations above the slope in the vicinity of such a boundary has to be replaced by a gradient (determined in PAX).

The mean nodal coolant temperature or, if L_{FTYPE}=0, void fraction of the last SC node is interrelated by the PAX procedure with the locally varying SC outlet boundary z_{CA}. Hence, in a transient situation the time-derivative of only one of these parameters is demanded. The second one follows then from the PAX procedure after the integration.

If combining (in case of single-phase flow) the eqs.(23) and (31), the wanted relation for the SC boundary time derivative can be expressed by

$$\frac{d}{dt}z_{CA}=\frac{d}{dt}z_{BB}=\frac{T^{t}_{PAXCA}-T^{t}_{TCA}}{T^{z}_{TCA}-T^{z}_{PAXCA}} \text{ or } =0 \qquad \left(n=N_{CT},\, z_{CA}<z_{BA} \text{ or } z_{CA}=z_{BA} \text{ if } L_{FTYPE}>0\right) \tag{66}$$

and if taking for the case of a mixture flow the eqs. (23) and (48) into account

$$\frac{d}{dt}z_{CA}=\frac{d}{dt}z_{ML}=\frac{\alpha^{t}_{PAXCA}-\alpha^{t}_{ACA}}{\alpha^{z}_{CA}\,\alpha^{z}_{PAXCA}} \text{ or } =0 \qquad \left(n=N_{CT},\, z_{CA}<z_{BA} \text{ or } z_{CA}=z_{BA} \text{ if } L_{FTYPE}=0\right) \tag{67}$$

If $z_{CA} < z_{BA}$, the corresponding time-derivatives $\frac{d}{dt}T_{Mn}$ or $\frac{d}{dt}\alpha_{Mn}$ of the last SC node (at n=N_{CT}) follow by inserting the terms above into the eqs.(31) or (48). After the integration procedure then the SC outlet boundary z_{CA} (= boiling boundary z_{BB} or mixture level z_{ML}) and thus also the total number N_{CT} of SC nodes are given.

3.10. Pressure profile along a SC (and thus also BC)

After having solved the mass and energy balance equations, separately and not simultaneously with the momentum balance, the now exact nodal SC and BC pressure difference terms ($\Delta P_{Nn} = P_{Nn} - P_{Nn-1}$ and ΔP_{BNn}) can be determined for both single- or two-phase flow situations by discretizing the momentum balance eq.(3) and, if applying a modified 'finite element method', integrating the eqs.(4) to (6) over the corresponding SC nodes. The total BC pressure difference $\Delta P_{BT} = P_{BA} - P_{BE}$ between BC outlet and entrance follows then from the relation

$$\Delta P_{BT} = \Delta P_{PBT} - \Delta P_{GBT} \qquad (\text{with } \Delta P_{GBT,0} = 0 \text{ at steady state conditions}) \tag{68}$$

with the part

$$\begin{aligned} \Delta P_{PBT} &= \Delta P_{SBT} + \Delta P_{ABT} + \Delta P_{XBT} + \Delta P_{FBT} + \Delta P_{DBT} \\ &\qquad (\text{with } \Delta P_{PBT,0} = \Delta P_{BTIN,0} \text{ at steady state conditions}) \end{aligned} \tag{69}$$

comprising, as described in section 2.1.3, terms from static head (ΔP_{SBT}), mass acceleration (ΔP_{ABT}), wall friction (ΔP_{FBT}) and external pressure accelerations (ΔP_{XBT} due to pumps or other perturbations from outside) and an (only in the transient situation needed) term ΔP_{GBT}, being represented as

$$\begin{aligned} \Delta P_{GBT} &= \int_0^{z_{BT}} \frac{d}{dt} G_{FB}(z,t)\, dz = z_{BT} \frac{d}{dt} G_{FBMT} && \text{at transient conditions} \\ &= 0 && \text{at steady state} \end{aligned} \tag{70}$$

This last term describes the influence of time-dependent changes in total mass flux along a BC (caused by the direct influence of changing nodal mass fluxes) and can be estimated by introducing a 'fictive' mean mass flux term G_{FBMT} (averaged over the entire BC)

$$\begin{aligned} G_{FBMT} &= \frac{1}{z_{BT}} \int_0^{z_{BT}} \frac{d}{dt} G_{FB}(z,t)\, dz \\ &\cong \frac{1}{z_{BT}} \sum^{N_{SCT}} \sum_{n=1}^{N_{CT}} \Delta z_{Nn} G_{FBMn} = \frac{1}{2} \frac{1}{z_{BT}} \sum_{1}^{N_{BT}} \frac{\Delta z_{Bk}}{A_{BMk}} (G_{Bk} + G_{BK-1}) \end{aligned} \tag{71}$$

Its time derivative can then be represented by

$$\begin{aligned} \frac{d}{dt} G_{FBMT} &\cong \frac{G_{FBMT} - G_{FBMTB}}{\Delta t} \\ &\left(\frac{d}{dt} G_{FBMT}\right) \quad \text{if } \Delta t = t - t_B \;\; 0 \;\; (\text{Index B = begin of time-step}) \end{aligned} \tag{72}$$

http://dx.doi.org/10.5772/53372

Looking at the available friction correlations, there arises the problem how to consider correctly contributions from spacers, tube bends, abrupt changes in cross sections etc. as well. The entire friction pressure decrease (ΔP_{FBT}) along a BC can thus never be described in a satisfactory manner solely by analytical expressions. To minimize these uncertainties a further friction term, ΔP_{DBT}, had to be included into these considerations.

The corresponding steady state additive pressure difference term $\Delta P_{DBT,0}$ is (according to the eqs.(69) and (70)) fixed since it can be assumed that the total steady state BC pressure difference $\Delta P_{BT,0} = \Delta P_{PBT,0}$ is known from input. It seems, however, to be reasonable to treat this term as a 'friction' (or at least the sum with ΔP_{FBT}) and not as a 'driving' force. Thus it must be demanded that these terms should remain negative. Otherwise, input terms such as the entire pressure difference along the BC or corresponding friction factors have to be adjusted in an adequate way.

Assuming the general additive pressure difference term ΔP_{DBT} may have the form

$$\Delta P_{DBT} = \left(f_{FMP,0} - 1\right)\Delta P_{FBT} + \Delta P_{FADD} \tag{73}$$

means that ΔP_{DBT} is either supplemented with a direct additive term (index FADD) or the friction part is provided with a multiplicative factor ($f_{FMP,0}$-1). For the additive part it will be assumed to be the (1-ε_{DPZ})-th part of the total additional pressure difference term and to be proportional to the square of the total coolant mass flow, e.g., at BC entrance

$$\Delta P_{FADD} = -\, f_{ADD} z_{BT} \left(\frac{G_F|G_F|}{2\varrho d_H}\right)_{BE} = (1 - \varepsilon_{DPZ})\Delta P_{DBT} \text{ and} = 0 \text{ (i.e., } \varepsilon_{DPZ}=1) \text{ if } \Delta P_{DBT,0} > 0 \tag{74}$$

with the input coefficient $\varepsilon_{DPZ} = \varepsilon_{DPZI}$ governing from outside which of them should prevail.

From the now known steady state total 'additional' term $\Delta P_{DBT,0}$ the corresponding additive friction factor $f_{ADD,0}$ follows then directly from the equation above, from eq.(73) then the multiplicative one

$$f_{FMP,0} = 1 + \varepsilon_{DPZ} \frac{\Delta P_{DBT,0}}{\Delta P_{FBT,0}} \text{(setting } \varepsilon_{DPZ}=1 \text{ if } \Delta P_{DBT,0} > 0) \tag{75}$$

For the steady state description of a 3D compartment with (identical) parallel channels in a first step a factor should be derived for a 'mean' channel. Then this factor can be assumed to be valid for all the other (identical) parallel channels too (See Jewer et al., 2005).
For the transient case there arises the question how the validity of both friction factors could be expanded to this situation too. In the here presented approach this is done by assuming that these factors should remain time-independent, i.e., that $f_{ADD} = f_{ADD,0}$ and $f_{FMP} = f_{FMP,0}$. This allows finally also to determine the wanted nodal pressure decrease term ΔP_{DBT} of eq.(69) for the transient case.

By adding the resulting nodal BC pressure difference terms to the (time-varying) system pressure $P_{SYS}(t)$, given from outside as boundary condition with respect to a certain position (in- or outside of the BC), then finally also the absolute nodal pressure profile P_{Bk} along the BC can be established (This term is needed at the begin of the next time step within the constitutive equations).

3.11. BC entrance mass flow ('Open and closed channel concept')

As to be seen from the sections above in order to be able to calculate the characteristic nodal and total single- and two-phase parameters along a BC the BC entrance mass flow must be known. This term is for the steady state case given by input ($G_{BE} = G_{BEIN}$), together with the two pressure entrance and outlet values ($P_{BE} = P_{BEIN}$, $P_{BA} = P_{BAIN}$). In the transient situation it can, however, be expected that for the normal case of an 'open' channel besides the entrance mass flow only one of these two pressure terms is known by input, either at BC entrance or outlet. The missing one follows then from the calculation of the pressure decrease part.

Such a procedure can not be applied without problems if the channels are part of a complex set of closed loops. Loops consisting of more than one coolant channel (and not driven by an outside source, such as a pump). Then the mass flow values (and especially the entrance term of at least one of the channels) has to be adjusted with respect to the fact that the sum of the entire pressure decrease terms along such a closed circuit must be zero. Usually, in the common thermal-hydraulic codes (see for example the 'separate-phase' approaches) this problem is handled by solving the three (or more) fundamental equations for the entire complex system simultaneously, a procedure which affords very often immense computational times and costs. In the here applied module (based on a separate treatment of momentum from mass and energy balance) a more elegant method could be found by introducing an additional aspect into the theory of CCM. It allows, different to other approaches, to take care of this situation by solving this problem by means of a 'closed channel concept' (in contrast to the usual 'open channel' method).

Choosing for this purpose a characteristic 'closed' channel within such a complex loop it can be expected that its pressure difference term $\Delta P_{BT} = \Delta P_{BA} - \Delta P_{BE}$ over this channel is fix (= negative sum of all the other decrease terms of the remaining channels which can be calculated by the usual methods). Thus also the outlet and entrance BC pressure values (P_{BAIN}, ΔP_{BEIN}) are now available as inputs to CCM. Since, according to eq.(69), the term ΔP_{PBT} is known too, then from eq.(68) a 'closed channel concept criterion' can be formulated

$$z_{BT} \frac{d}{dt} G_{FBMT} = \Delta P_{GBT} = \Delta P_{PBT} - \Delta P_{BTIN} \quad \text{(at 'closed channel' conditions)} \tag{76}$$

demanding that in order to fulfil this criterion the total mass flow along a BC (and thus also at its entrance) must be adapted in such a way that the above criterion which is essential for the 'closed channel' concept remains valid at each time step, i.e., that the actual time derivative of the mean mass flux G_{FBMT} averaged over the channel must, as recommended by eq.(71), agree in a satisfactory manner with the required one from the equation above.

There exist, obviously, different methods how to deal with this complicated problem. One of them could be to determine the entrance mass flow G_{BE} by changing this value in a recursive way until the resulting term $\frac{d}{dt}G_{FBMT}$ agrees with the criterion above.

Another possibility is to find a relation between the time-derivatives of the mean and of special local mass flux values (e.g., at BC entrance), i.e., to establish for example a relation between the terms $\frac{d}{dt}G_{FBMT}$ and $\frac{d}{dt}G_{FBE}$. Then the wanted mass flow time-derivative at BC entrance can be determined directly from eq.(76). One practicable method to establish such a relation could follow if considering that a change of the mass flux is propagating along the channel so fast that the time derivative of its mean values could be set (in a first step) almost equal to the time-derivative at its entrance value. This term can, eventually, be provided with a form factor which can be adapted by an adequate recursion procedure until the condition of eq.(75) is fulfilled. The so won entrance mass flow is then governing the mass flow behaviour of the entire loop.

This last and very provisional method has been tested by applying the UTSG-3 code (Hoeld, 2011b) for the simulation of the natural-circulation behaviour of the secondary steam generator loop. Similar considerations have been undertaken for a 3D case where the automatic mass flow distribution into different entrances of a set of parallel channels had to be found (See e.g. Hoeld, 2004a and Jewer et al., 2005). The experience of such calculations should help to decide which of the different possible procedures should finally be given preference.

The 'open/closed channel concept' makes sure that measures with regard to the entire closed loop do not need to be taken into account simultaneously but (for each channel) separately. Its application can be restricted to only one 'characteristic' channel of a sequence of channels of a complex loop. This additional tool of CCM can in such cases help to handle the variety of closed loops within a complex physical system in a very comfortable way.

4. Code package CCM

Starting from the above presented 'drift-flux based mixture-fluid theory' a (1D) thermal-hydraulic coolant channel module, named CCM, could be established. A module which was derived with the intention to provide the authors of different and sometimes very complex thermal-hydraulic codes with a general and easily applicable tool needed for the simulation of the steady state and transient behaviours of the most important single- and two-phase parameters along any type of heated or cooled coolant channel.

For more details about the construction and application of the module CCM see (Hoeld, 2005, 2007a and b and 2011b).

5. Verification and Validation (V & V) procedures

During the course of development of the different versions of the code combination UTSG-3/ CCM the module has gone through an appropriate verification and validation (V&V) procedure (with continuous feedbacks being considered in the continual formulation of the theoretical model).

CCM is (similar as done in the separate-phase models) constructed with the objective to be used only as an element within an overall code. Hence, further V&V steps could be performed only in an indirect way, i.e. in combination with such overall codes. This has been done in a very successful way by means of the U-tube steam generator code UTSG-3. Thereby the module CCM could profit from the experiences been gained in decades of years work with the construction of an effective non-linear one-dimensional theoretical model and, based on it, corresponding digital code UTSG-2 for vertical, natural-circulation U-tube steam generators (Hoeld, 1978, 1988b, 1990a and 1990b), (Bencik et al., 1991) and now also the new advanced code version UTSG-3 (Hoeld 2002b, 2004a).

The good agreement of the test calculations with similar calculations of earlier versions applied to the same transient cases demonstrates that despite of the continuous improvements of the code UTSG and the incorporation of CCM into UTSG-3 the newest and advanced version has still preserved its validity.

A more detailed description over these general V&V measures demonstrated on one characteristic test case can be found in (Hoeld, 2011b).

6. Conclusions

The universally applicable coolant channel module CCM allows describing the thermal-hydraulic situation of fluids flowing along up-, horizontal or downwards channels with fluids changing between sub-cooled, saturated and superheated conditions. It must be recognized that CCM represents a complete system in its own right, which requires only BC-related, and thus easily available input values (geometry data, initial and boundary conditions, resulting parameters from integration). The partitioning of a basic channel into SC-s is done automatically within the module, requiring no special actions on the part of the user. At the end of a time-step the characteristic parameters of all SC-s are transferred to the corresponding BC positions, thus yielding the final set of ODE-s together with the parameters following from the constitutive equations of CCM.

In contrast to the currently very dominant separate-phase models, the existing theoretical inconsistencies in describing a two-phase fluid flowing along a coolant channel if changing between single-phase and two-phase conditions and vice versa can be circumvented in a very elegant way in the 'separate-region' mixture-fluid approach presented here. A very unique technique has been established built on the concept of subdividing a basic channel (BC) into different sub-channels (SC-s), thus yielding exact solutions of the basic drift-flux supported conservation equations. This type of approach shows, as discussed in (Hoeld, 2004b), distinct advantages vs. 'separate phase' codes, especially if taking into account

- the quality of the fundamental equations (basic conservation equations following directly from physical laws supported by experimentally based constitutive equations vs. split 'field' equations with artificial closure terms),
- the special solution methods due to the detailed interpolation procedure from PAX allowing to calculate the exact movement of boiling boundaries and mixture (or dry-out) levels (different to the 'donor-cell averaging' methods yielding mostly only 'condensed' levels),
- the easy replacement of new and improved correlations within the different packages without having to change the basic equations of the theory (for example the complicated exchange terms of a 'separate-phase' approach),
- the possibility to take advantage of the 'closed-channel concept' (needed for example for thermal- hydraulic 3D considerations) allowing thus to decouple a characteristic ('closed') channel from other parts of a complex system of loops,
- the speed of the computation,
- the derivation of the theory in close connection with the establishment of the code by taking advantage of feedbacks coming from both sides,
- the considerable effort that has been made in verifying and checking CCM (besides an extensive V & V procedure), with respect to the applicability and adjustment and also for very extreme situations,
- its easy applicability,
- the maturity of the module, which is continuously enhanced by new application cases,
- taking advantage of the fact that most of the development work for the coolant channel thermal-hydraulics has already been shifted to the establishment of the here presented module (including the special provisions for extreme situations such as stagnant flow, zero power or zero sub-cooling, test calculations for the verification and validation of the code etc.).

The existence of the resulting widely verified and validated module CCM represents an important basic element for the construction of a variety of other comprehensive thermal-hydraulic models and codes as well. Such models and modules can be needed for the simulation of the steady state and transient behaviour of different types of steam generators, of 3D thermal-hydraulic compartments consisting of a number of parallel channels (reactor cores, VVER steam generators etc.). It shows special advantages in view of the determination of the mass flow distribution into different coolant channels after non-symmetric perturbations see (Hoeld, 2004a) or (Jewer et al., 2005), a problem which is far from being solved in many of the newest 3D studies.

The introduction of varying cross sections along the z axis allows to take care also of thermal non-equilibrium situations by simulating the two separate phases by two with each other interacting basic channels (for example if injecting sub-cooled water rays into a steam dome).

The resulting equations for different channels appearing in a complex physical system can be combined with other sets of algebraic equations and ODE-s coming from additional parts of such a complex model (heat transfer or nuclear kinetics considerations, top plenum, main steam system and downcomer of a steam generator etc.). The final overall set of ODE-s can then be solved by applying an appropriate time-integration routine (Hoeld, 2004a, 2005, 2007a and b, 2011b).

The enormous efforts already made in the verification and validation of the codes UTSG-3, its application in a number of transient calculations at very extreme situations (fast opening of safety valves, dry out of the total channel with SC-s disappearing or created anew) brings the code and thus also CCM to a very mature and (what is important) easily applicable state. However, there is not yet enough experience to judge how the potential of the mixture-fluid models and especially of CCM can be expanded to other extreme cases (e.g., water and steam hammer). Is it justified to prefer separate-phase models versus the drift-flux based (and thus non-homogeneous) mixture fluid models? This depends, among other criteria, also on the quality of the special models and their exact derivation. Considering the arguments presented above it can, however, be stated that in general the here presented module can be judged as a very satisfactory approach.

7. Nomenclature

A_{Bk}	m^2	BC cross sectional area (at BC node boundary k)
A_{Nn}, A_{Mn}	m^2	SC cross sectional area (at SC node boundary n, mean value)
A_{TWMn}	m^2	Surface area of a (single) tube wall along a node n
C	-	Dimensionless constant
C_0	-	Phase distribution parameter
d_{HY}	m	Hydraulic diameter
$f(z,t)$, f_{Nn}, f_{Mn}	-	General and nodal (boundary and mean) solution functions
f_{LIMCA}	-	Upper or lower limit of the approx. function f(z,t)
f_{ADDO}, f_{FMPO}	-	Additive and multiplicative friction coefficients
$G = G_W + G_S$ $=A[(1-\alpha)\rho_W v_W + \alpha v_S \rho_S]$	$\frac{kg}{s}$	Mass flow
$GF = \frac{G}{A} = v\,\rho$	$\frac{kg}{sm^2}$	Mass flux
h, h^P, $c_P = h^T$	$\frac{J}{kg}, \frac{m^3}{kg}, \frac{J}{m^3 kg}$	Specific enthalpy and its partial derivatives with respect to pressure and temperature (= specific heat)
$h_{SW} = h'' - h'$, h'', h'	$\frac{J}{kg}$	Latent heat, saturation steam and water enthalpy
K_{EYBC}	-	Characteristic key number of channel BC
L_{FTYPE}= 0, 1 or 2	-	SC with saturated water/steam mixture, sub-cooled water or superheated steam
L_{FTBE} (=L_{FTYPE} of 1-st N_{SC})	-	L_{FTYPE} of 1-st SC within BC

N_{BT}	-	Total number of BC nodes
$N_{BCA} = N_{CT}+N_{BCE}-1$, N_{BCE}	-	BC node numbers containing SC outlet or entrance
$N_{CT}=N_{BCA}-N_{BCE}+1$	-	Total number of SC nodes
$N_{SC} = N_{SCE}$, N_{SCA}	-	Characteristic number of SC, setting N_{SCE}= 1, 2 or 3 if L_{FTYPE}= 0, 1 or 2 and $N_{SCA}=N_{SCE}+N_{SCT}-1$
P, $\Delta P_T = P_A - P_E$	$Pa=\frac{J}{m^3}=\frac{kg}{ms^2}$	Pressure and pressure difference (in flow direction)
Q_{BT}, Q_{BMk}	W	Total and nodal power into BC node k
$q_{BMk}=\frac{U_{Bk}}{A_{BMk}}q_{LBMk}$	$\frac{W}{m^3}$	Mean nodal BC power density into the fluid (= volumetric heat transfer rate)
$q_{FBk}=A_{Bk}\frac{q_{Bk}}{U_{Bk}}=\frac{q_{LBk}}{U_{TWBk}}$	$\frac{W}{m^2}$	Heat flux from (heated) wall to fluid at BC node k
$q_{LBk}=\frac{Q_{Bk}}{\Delta z_{Bk}}=A_{Bk}q_{Bk}$	$\frac{W}{m}$	Linear power along BC node k
$q_{Mn}=\frac{Q_{Mn}}{V_{Mn}} = \frac{1}{2A_{Mn}}(q_{LNn}+q_{LNn-1})$	$\frac{W}{m^3}$	Mean nodal SC power density into fluid (= volumetric0 heat transfer rate)
T, t	C, s	Temperature, time
U_{TW}	m	Perimeter of a heated (single) tube wall
$V_{Mn}=\frac{1}{2}(A_{Nn}+A_{Nn-1})\Delta z_{Nn}$	m^3	Mean nodal SC volume
$v_S = \frac{G_{FS}}{\alpha\rho_S}$, $v_W = \frac{G_{FW}}{(1-\alpha)\rho_W}$	$\frac{m}{s}$	Steam and water velocity
$X=\frac{G_S}{G}$ or $=\frac{h-h'}{h_{SW}}$	-	Steam quality (2- resp. extended to 1-phase flow)
z, $\Delta z_{Nn}=z_{Nn}-z_{Nn-1}$	m	Local position, SC node length ($z_{Nn-1}=z_{CE}$ at n=0)
$z_{BA}-z_{BE}=z_{BT}$, $z_{CA}-z_{CE}=z_{CT}$	m	BC and SC outlet and entrance positions, total length
z_{BB}, z_{ML}, z_{SPC}	m	Boiling boundary, mixture level resp. supercritical boundary within a BC
α	-	Void fraction
α_{TWk}	$\frac{W}{m^2C}$	Heat transfer coefficient along a BC wall surface
Δ	-	Nodal differences
ε_{DPZ}	-	Coefficient controlling the additional friction part
ε_{QTW}	-	Correction factor with respect to $Q_{NOM,0}$
ε_{TW}	m	Abs. roughness of tube wall (ε_{TW}/d_{HW}= relative value)
Φ^2_{2PF}	-	Two-phase multiplier
Φ_{ZG}	-	Angle between upwards and flow direction
ρ, ρ^P, ρ^T	$\frac{kg}{m^3}$, $\frac{kg}{J}$, $\frac{kg}{m^3C}$	Density and their partial derivatives with respect to

		(system) pressure and temperature
Θ	s	Time constant
∂	-	Partial derivative

Subscripts

0, 0 (=E or BE)	Steady state or entrance to SC or BC (n or k =0)
A, E, T (=AE)	Outlet, entrance, total (i.e. from outlet to entrance)
B, S	Basic channel or sub-channel (=channel region)
A, S, F, D, X (P=A+S+F+D+X) and G	Acceleration, static head, direct and additional Friction, external pressure differences (in connection with ΔP) and pressure differences due to changes in mass flux
Mn, BMk	Mean values over SC or BC nodes
Nn, Bk	SC or BC node boundaries (n = 0 or k = 0: Entrance)
D	Drift
S, W	Steam, water
P, T	Derivative at constant pressure or temperature
TW	Tube wall surface

Superscripts

/,//	Saturated water or steam
P, T	Partial derivatives with respect to P or T
(G_S), (α), z, s	Partial derivatives with respect to G_S, .α. or z (=gradient), slope

Acronyms

ATHLET, CATHARE; CATHENA, RELAP, TRAC	Well-known thermal-hydraulic codes (on the basis a separate-phase approach)
BC, SC	Basic(=coolant) channel subdivided into subchannels (=regions of different flow types)
BWR, PWR	Boiling and pressurized reactors
CCF	Counter-current flow
CCM	Coolant channel model and module (on the basis a separate-region approach)

http://dx.doi.org/10.5772/53372

GCSM	General Control Simulation Module (Part of ATHLET)
GRS	Gesellschaft für Anlagen- und Reaktorsicherheit
HETRAC	Heat transfer coefficients package
HTC	Heat transfer coefficients
MDS	Drift flux code package
MPPWS, MPPETA	Thermodynamic and transport property package
NPP	Nuclear power plant
PAX (PAXDRI)	Quadratic polygon approximation procedure (with driver code)
PDE, ODE	Partial and ordinary differential equation
UTSG	U-tube steam generator code
V & V	Verification and validation procedure

Author details

Alois Hoeld*

Retired from GRS, Garching/Munich, Munich, Germany

References

[1] Austregesilo, H, et al. (2003). *ATHLET Mod 2.0, Cycle A, Models and methods*. GRS-P-1/ Vol.4.

[2] Bencik, V, & Hoeld, A. (1991). *Experiences in the validation of large advanced modular system codes.* 1991 SCS Simulation MultiConf., April 1-5New Orleans

[3] Bencik, V, & Hoeld, A. (1993). *Steam collector and main steam system in a multi-loop PWR NPP representation.* Simulation MultiConf., March 29 April 1, Arlington/Washington

[4] Bestion, D. (1990). *The physical closure laws in CATHARE code*. Nucl.Eng.& Design Burwell, J. et al. 1989. *The thermohydraulic code ATHLET for analysis of PWR and SWR system,* Proc.of 4-th Int. Topical Meeting on Nuclear Reactor Thermal-Hydraulics, NURETH-4, Oct.10-13, Karlsruhe, 124, 229-245.

[5] Fabic, S. (1996). *How good are thermal-hydraulics codes for analyses of plant transients* International ENS/HND Conference, Oct. 1996, Proc.193-201, Opatija/Croatia, 7-9.

[6] Haar, L, Gallagher, J. S, & Kell, G. S. (1988). *NBS/NRC Wasserdampftafeln*. London/Paris/ New York 1988

[7] Hanna, B. N. (1998). *CATENA. A thermalhydraulic code for CANDU analysis.* Nucl. Eng. Des. , 180(1998), 113-131.

[8] Hoeld, A. (1978). *A theoretical model for the calculation of large transients in nuclear natural circulation U-tube steam generators (Digital code UTSG).* Nucl. Eng. Des. (1978), 47, 1.

[9] Hoeld, A. *HETRACA heat transfer coefficients package.* GRS-A-1446, May.

[10] Hoeld, A. (1988b). *Calculation of the time behavior of PWR NPP during a loss of a feedwater ATWS case.* NUCSAFE October, Proc. 1037-1047, Avignon, France., 88, 2-7.

[11] Hoeld, A. *UTSG-2. A theoretical model describing the transient behaviour of a PWR natural-circulation U-tube steam generator*. Nuclear Techn. April., 90, 98-118.

[12] Hoeld, A. (1990b). *The code ATHLET as a valuable tool for the analysis of transients with and without SCRAM*, SCS Eastern MultiConf,April, Nashville, Tennessee., 23-26.

[13] Hoeld, A. (1996). *Thermodynamisches Stoffwertpaket MPP für Wasser und Dampf.* GRS, Technische Notiz, TN-HOD-1/96, Mai.

[14] Hoeld, A. (1999). *An Advanced Natural-Circulation U-tube Steam Generator Model and Code UTSG-3*, EUROTHERM-SEMINAR September Genoa, Italy.(63), 6-8.

[15] Hoeld, A. (2000). *The module CCM for the simulation of the thermal-hydraulic situation within a coolant channel*. Intern. NSS/ENS Conference, September Bled, Slovenia., 11-14.

[16] Hoeld, A. (2001). *The drift-flux correlation package MDS*. th International Conference on Nuclear Engineering (ICONE-9), 8-12 April, Nice, France., 9.

[17] Hoeld, A. (2002a). *The consideration of entrainment in the drift-flux correlation package MDS*. th Int. Conference on Nuclear Engineering (ICONE-10), April 14-18, Arlington, USA., 10.

[18] Hoeld, A. (2002b). *A steam generator code on the basis of the general coolant channel module CCM*. PHYSOR 2002, October Seoul, Korea., 7-10.

[19] Hoeld, A. (2004a). *A theoretical concept for a thermal-hydraulic 3D parallel channel core model.* PHYSOR 2004, April Chicago, USA., 25-29.

[20] Hoeld, A. (2004b). *Are separate-phase thermal-hydraulic models better than mixture-fluid approaches. It depends. Rather not*. Int. Conference on Nuclear Engineering for New Europe 2004'. Sept. Portoroz, Slovenia., 6-9.

[21] Hoeld, A. *A thermal-hydraulic drift-flux based mixture-fluid model for the description of single- and two-phase flow along a general coolant channel.* The th International Topical Meeting on Nuclear Reactor Thermal-Hydraulics (NURETH-11). Paper 144. Oct. 2-6, 2005, Avignon, France., 11.

[22] Hoeld, A. (2007a). *Coolant channel module CCM. An universally applicable thermal-hydraulic drift-flux based mixture-fluid 1D model and code.* Nucl. Eng. and Desg. 237(2007), 1952-1967

[23] Hoeld, A. (2007b). *Application of the mixture-fluid channel element CCM within the U-tube steam generator code UTSG-3 for transient thermal-hydraulic calculations.*th Int. Conf. on Nuclear Engineering (ICONE-15). Paper 10206. April 22-26, 2007, Nagoya, Japan, 15.

[24] Hoeld, A. (2011a). *Coolant channel module CCM. An universally applicable thermal-hydraulic drift-flux based separate-region mixture-fluid model.* Intech Open Access Book 'Steam Generator Systems: Operational Reliability and Efficiency'', V. Uchanin (Ed.), March 2011. 978-9-53307-303-3InTech, Austria, Available as: http://www.intechopen.com/articles/show/title/*the-thermal-hydraulic- u-tube-steam-generator- model-and-code-utsg-3-based-on-the-universally- applicab* or http://www.intechopen.com/search?q=hoeld

[25] Hoeld, A. (2011b). *The thermal-hydraulic U-tube steam generator model and code UTSG3 (Based on the universally applicable coolant channel module CCM).* See Intech Open Access Book referred above.

[26] Hoeld, A, & Jakubowska, E. Miro, J.E & Sonnenburg,H.G. (1992). *A comparative study on slip and drift- flux correlations.* GRS-A-1879, Jan. 1992

[27] Ishii, M. (1990). *Two-Fluid Model for Two-Phase Flow,* Multiphase Sci. and Techn. 5.1, Hemisphere Publ. Corp.

[28] Ishii, M, & Mishima, K. (1980). *Study of two-fluid model and interfacial area.* NUREG/CR-1873 (ANL-80-111), Dec.

[29] Jewer, S, Beeley, P. A, Thompson, A, & Hoeld, A. (2005). *Initial version of an integrated thermal-hydraulic and neutron kinetics 3D code X3D.* ICONE13, May Beijing. See also NED 236(2006), 1533-1546., 16-20.

[30] Lerchl, G, et al. (2009). *ATHLET Mod 2.2, Cycle B, User's Manual'.* GRS-Rev. 5, July, 1

[31] Liles, D. R, et al. (1988). *TRAC-PF1/MOD1 Correlations and Models.* NUREG/CR-5069, LA-11208-MS, Dec.

[32] Martinelli, R. C, & Nelson, D. B. (1948). *Prediction of pressure drop during forced-circulation of boiling water.* Trans. ASME 70, 695.

[33] Moody, N. L. F. (1994). *Friction factors for pipe flow.* Trans. ASME, 66, 671 (See also VDI-Wärmeatlas, 7.Auflage, 1994, VDI- Verlag).

[34] Schmidt, E, & Grigull, U. ed.) (1982). *Properties of water and steam in SI-Units.* Springer-Verlag.

[35] Sonnenburg, H. G. (1989). *Full-range drift-flux model based on the combination of drift-flux theory with envelope theory,* NURETH-4, Oct.10-13, Karlsruhe., 1003-1009.

[36] Shultz, R. R. (2003). *RELAP5/3-D code manual. Volume V: User's guidelines,* INEEL-EXT-98-00084. Rev.2.2, October.

[37] , U. S-N. R. C. 2001a. *TRAC-M FORTRAN90 (Version 3.0). Theory Manual,* NUREG/CR-6724, April

[38] , U. S-N. R. C. 2001b. *RELAP5/MOD3.3. Code Manual.* Volume I, NUREG/CR-5535, December

[39] Zuber, N, & Findlay, J. A. (1965). Average volume concentration in two-phase flow systems, J. Heat Transfer 87, 453

Chapter 2

Large Eddy Simulation of Thermo-Hydraulic Mixing in a T-Junction

Aleksandr V. Obabko, Paul F. Fischer,
Timothy J. Tautges, Vasily M. Goloviznin,
Mikhail A. Zaytsev, Vladimir V. Chudanov,
Valeriy A. Pervichko, Anna E. Aksenova and
Sergey A. Karabasov

Additional information is available at the end of the chapter

http://dx.doi.org/10.5772/53143

1. Introduction

Unsteady heat transfer problems that are associated with non-isothermal flow mixing in pipe flows have long been the topic of concern in the nuclear engineering community because of the relation to thermal fatigue of nuclear power plant pipe systems. When turbulent flow streams of different velocity and density rapidly mix at the right angle, a contact interface between the two mixing streams oscillates and breaks down because of hydrodynamic instabilities, and large-scale unsteady flow structures emerge (Figure 1). These structures lead to low-frequency oscillations at the scale of the pipe diameter, D, with a period scaling as $O(D/U)$, where U is the characteristic flow velocity. If the mixing flow streams are of different temperatures, the hydrodynamic oscillations are accompanied by thermal fluctuations (thermal striping) on the pipe wall. The latter accelerate thermal-mechanical fatigue, damage the pipe structure and, ultimately, cause its failure.

The importance of this phenomenon prompted the nuclear energy modeling and simulation community to establish a common benchmark to test the ability of computational fluid dynamics (CFD) tools to predict the effect of thermal striping. The benchmark is based on the thermal and velocity data measurements obtained in a recent Vattenfall experiment designed specifically for this purpose [1, 2]. Because thermal striping is associated with large-scale anisotropic mixing, standard engineering modeling tools based on time-averaging approaches, such as Reynolds-averaged Navier-Stokes (RANS) models, are badly suited. Consequently, one must consider unsteady modeling methods such as unsteady RANS (URANS), parabolised stability equations (PSE), or large eddy simulation (LES). Among these choices, the LES approach is most generic and modeling-assumption free, unlike the PSE and URANS methods, which, for instance, assume the existence of a spectral gap between the large and the small scales. On the other hand, owing to recent advances in computing, LES methods are becoming an increasingly affordable tool.

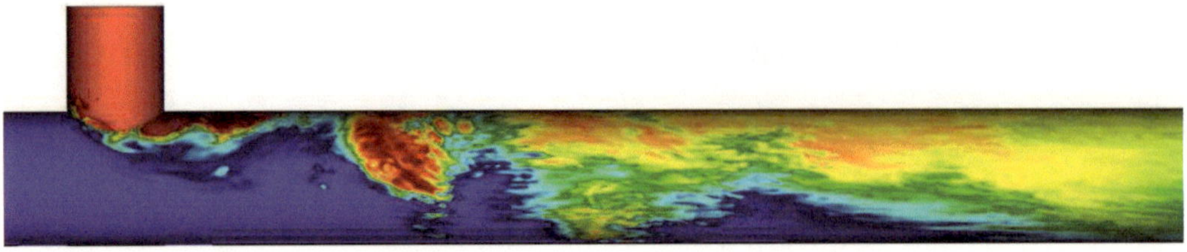

Figure 1. Instantaneous wall temperature distribution from Nek5000 T-junction simulation.

In this chapter, we compare the results of three LES codes produced for the Vattenfall benchmark problem: Nek5000, developed at Argonne National Laboratory in the United States, and CABARET and Conv3D, developed at the Moscow Institute for Nuclear Energy Safety (IBRAE) in Russia. Nek5000 is an open source code based on the spectral-element method (SEM), a high-order weighted residual technique that combines the geometric flexibility of the finite-element method with the tensor-product efficiencies of spectral methods [3, 4]. CABARET, which stands for Compact Accurately Boundary Adjusting high REsolution Technique, is based on the scheme of [5]. CABARET was developed as a significant upgrade of the second-order upwind leapfrog scheme for linear advection equation [6, 7] to nonlinear conservation laws in one and two dimensions [8–10]. The CABARET scheme is second-order accurate on nonuniform grids in space and time, is nondissipative, and is low-dispersive. For nonlinear flow calculations, it is equipped with a conservative nonlinear flux correction that is directly based on the maximum principle. For 3D calculations, a new unstructured CABARET solver was developed (e.g. [11, 12]) that operates mainly with hexagonal meshes. Conv3D is a well-established solver of IBRAE that has been validated on various experimental and benchmark data [13, 14]. For ease of the grid generation in the case of complex geometries, Conv3D utilizes the immersed boundary method (IBM) on Cartesian meshes with a cut-cell approach. The underlying scheme of Conv3D [15] is a low-dissipative, second-order approximation in space and a first-order approximation in time.

This chapter is organized as follows. In Section 2 we briefly describe the experimental setup and data collection procedure. In Sections 3–5 we outline the computational models. The results are discussed in Section 6, including a comparison of data from the simulations and experiment. Concluding remarks and a brief discussion of future work are provided in Section 7.

2. Experimental configuration

The Vatenfall experiment [16] is based on water flow in a main pipe of diameter D=140 mm with a side branch of diameter D_H=100 mm adjoining the main at a 90 degree angle. The pipes and the T-junction, which is made from a plexiglass block, are transparent so that the velocity can be measured with laser Doppler anemometry (LDA). Velocity data was taken under isothermal conditions with both flows entering at 19 °C. In order to measure thermal striping, time-dependent temperature data was collected from thermocouples downstream of the T-junction with flow at 19 °C entering the main branch and flow at 36 °C entering the side branch.

Parameter	Main Branch	Hot Branch
Diameter D^*, D^*_H (m)	0.1400	0.1000
Flow rate q^*, q^*_H (l/s)	9.00	6.00
Average velocity U^*, $U^*{}_H$ (m/s)	0.585	0.764
Inlet temperature (°C)	19.0	36.0
Density ρ (kg/m^3)	998.5	993.7
Dynamic viscosity $(N \times s/m^3)$	1.029e-3	7.063e-4
Kinematic viscosity ν (m^2/s)	1.031e-6	7.108e-7

Table 1. Dimensional Parameters for Vatenfall T-junction Experiment

Parameter	Main Branch	Hot Branch
Diameter D, D_H	1.000	0.714
Average velocity U, U_H	1.000	1.307
Inlet temperature	0.000	1.000
Density	1.000	0.9952
Reynolds number Re, Re_H	79400	107000

Table 2. Nondimensional Parameters for Vatenfall T-junction Experiment

The flow enters the cold (main) branch from a stagnation chamber located 80 D upstream of the T-junction and is assumed to be a fully developed turbulent flow by the time it reaches the T-junction. The hot branch flow enters at $20D_H$ upstream and is not quite fully developed as it enters the T-junction. The inlet flow rates are 9 and 6 liters per second (l/s), respectively in the cold and hot branches, which corresponds to a Reynolds number of $Re \approx 80,000$ and $100,000$, respectively. The key dimensional parameters are summarized in Table 1 and their nondimensional counterparts in Table 2. More detailed description of the T-junction benchmark experiment can be found in [1, 2].

3. Nek5000 simulations

The Nek5000 simulations are based on the spectral element method developed by Patera [3]. Nek5000 supports two formulations for spatial and temporal discretization of the Navier-Stokes equations. The first is the $\mathbb{P}_N - \mathbb{P}_{N-2}$ method [17–19], in which the velocity/pressure spaces are based on tensor-product polynomials of degree N and $N-2$, respectively, in the reference element $\hat{\Omega} := [-1,1]^d$, for $d = 2$ or 3. The computational domain consists of the union of E elements Ω^e, each of which is parametrically mapped from $\hat{\Omega}$ to yield a body-fitted mesh. The second is the low-Mach number formulation due to Tomboulides and Orszag [20, 21], which uses consistent order-N approximation spaces for both the velocity and pressure. The low-Mach number formulation is also valid in the zero-Mach (incompressible) limit [22]. Both formulations yield a decoupled set of elliptic problems to be solved at each timestep. In d=3 space dimensions, one has three Helmholtz solves of the form $(\beta I + \nu \Delta t A)\underline{u}^n_i = \underline{f}^n_i$, $i = 1,\ldots,d$, and a pressure Poisson solve of the form $A\underline{p}^n = \underline{g}^n$ at each timestep, t^n, $n = 1,\ldots$. Here, A is the symmetric positive-definite Laplace operator, and β is an order-unity coefficient coming from a third-order backward-difference approximation to the temporal derivative. (For the $\mathbb{P}_N - \mathbb{P}_{N-2}$ method, the Laplace operator A is replaced by a spectrally equivalent matrix arising from the unsteady Stokes equations [4, 19].) For marginally resolved LES cases, we find that the higher-order pressure approximation of the $\mathbb{P}_N - \mathbb{P}_N$ formulation tends to yield improved skin-friction estimates, and this is consequently the formulation considered here.

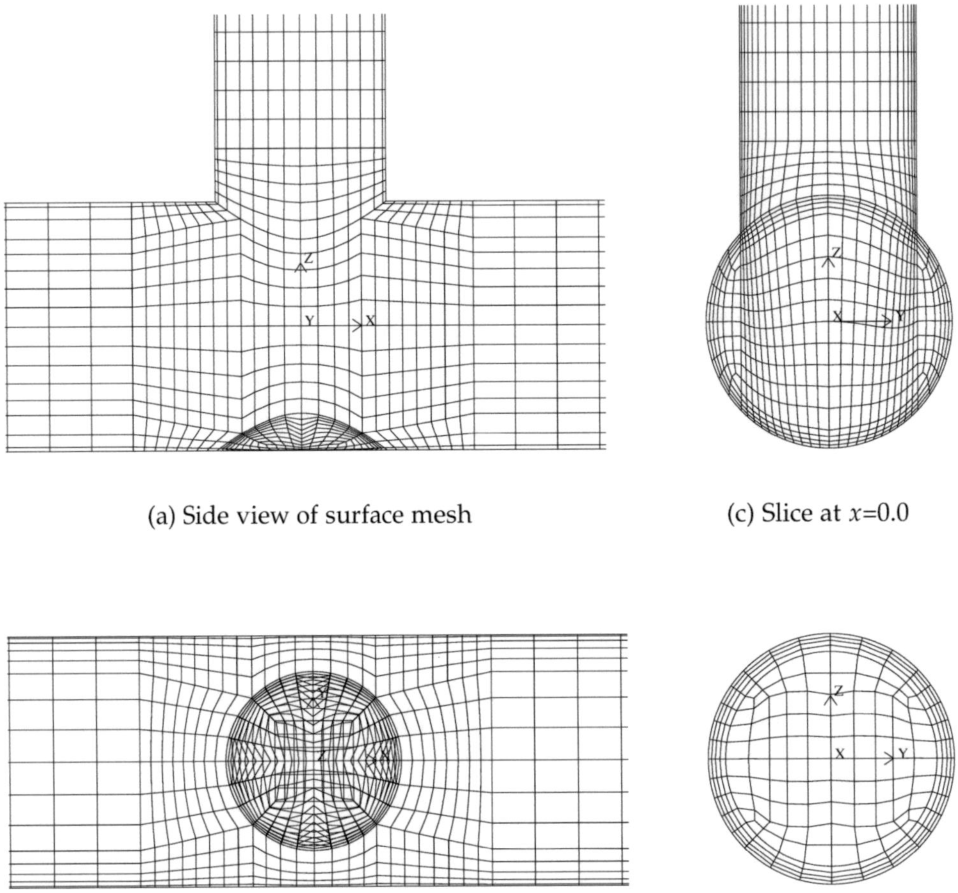

(a) Side view of surface mesh

(c) Slice at x=0.0

(b) Top view

(d) Slice at x=4.0

Figure 2. Computational mesh for Nek5000 T-junction simulation comprising E=62176 elements.

Consisting of $E = 62,176$ elements, the computational mesh for the Nek5000 simulations was generated by using CUBIT and read through the Nek-MOAB coupling interface [23]. Within each element, velocity and pressure are represented as Lagrange interpolating polynomials on tensor products of Nth-order Gauss-Lobatto-Legendre points. Unless otherwise noted, all the simulations were run with polynomial order N=7, corresponding to a total number of mesh points $n \approx EN^3 \approx 21$ million. Figure 2 shows a closeup of the mesh in the vicinity of the origin, which is located at the intersection of the branch centerlines. The inlet for the main branch is at x=-9.2143 and for the hot branch at z=6.4286. These lengths permitted generation of fully developed turbulence upstream of the T-junction, as described below. The outlet at x=22 was chosen to allow downstream tracking of temperature data at locations provided in the experiment. Away from the origin, the axial extent of the spectral elements in the main branch is 0.18 D, corresponding to a maximum axial mesh spacing of $\delta x_{\max} = 0.0377$. At the wall, the wall-normal element size is 0.01222, corresponding to a minimum spacing of $\delta y_n \approx 0.0008$. The submitted simulations were run with $Re = 40,000$ in the inlet branches,

http://dx.doi.org/10.5772/53143

yielding $Re = 60,000$ in the outlet. Downstream of the T-junction, the first grid point away from the wall is thus at $y^+ \approx 2.5$ in wall units.

Inlet flow conditions in the main branch are based on a recycling technique in which the inlet velocity at time t^n is given by $\alpha\mathbf{u}(\tilde{x}, y, z, t^{n-1})$, where $\tilde{x} = -3$ and α is chosen to ensure that the mass flow rate at inflow is constant ($\int u^n(-9, y, z)dy\,dz \equiv \pi/4$). Recycling is also used for the hot branch, save that the inflow condition is $0.8\beta\mathbf{u}(x, y, \tilde{z}, t^{n-1}) - 0.2U_H$, with β chosen so that the average inflow velocity is -U_H and $\tilde{z} = 2.1$. The 0.8 multiplier was added in order to give a flatter profile characteristic of the non-fully-developed flow realized in the experiment, but a systematic study of this parameter choice was not performed (see also Section 6.1.1).

Initial conditions for all branches were taken from fully developed turbulent pipe flow simulations at $Re = 40,000$. The timestep size was $\Delta t = 2.5 \times 10^{-4}$ in convective time units, or about 6×10^{-5} seconds in physical units corresponding to the experiment. The simulation was run to a time of t=28 convective time units prior to acquiring data. Data was then collected over the interval $t \in [28, 58]$ (in convective time units) for the benchmark submission[1] and over $t \in [28, 104]$ subsequently (longer average) both with $N = 7$. In addition to $N = 7$ results (i.e., with $n \approx EN^3 \approx 2.1 \times 10^7$ points), Nek5000 runs were conducted with $N = 5$ ($n \approx 7.7 \times 10^6$ points) and with $N = 9$ ($n \approx 4.5 \times 10^7$ points). The case with $N = 5$ started with the $N = 7$ results and was run and averaged over about 110 convective time units (i.e., about 26 seconds). The timestep size for $N = 5$ runs was twice as big (i.e, $\Delta t = 5 \times 10^{-4}$). The benchmark submission results required approximately 3.4×10^5 CPU hours on Intrepid (IBM BlueGene/P) while the follow-up study for a longer time average results took additional 5.9×10^5 CPU hours. Note that all Nek5000 results reported here were obtained with constant density equal to the nondimensional value of 1.000 (see Table 2).

4. CABARET simulations

The system of Navier-Stokes equations in a slightly compressible form (i.e., for a constant sound speed) is solved with a new code based on CABARET. In order to lessen the computational grid requirements, the code uses hybrid unstructured hexahedral/tetrahedral grids. CABARET is an extension of the original second-order, upwind leapfrog scheme [6] to nonlinear conservation laws [5, 24] and to multiple dimensions [9, 25]. In summary, CABARET is an explicit, nondissipative, conservative finite-difference scheme of second-order approximation in space and time. In addition to having low numerical dispersion, CABARET has a very compact computational stencil because of the use of separate variables for fluxes and conservation. The stencil is staggered in space and time and for advection includes only one computational cell. For nonlinear flows, CABARET uses a low-dissipative, conservative flux correction method directly based on the maximum principle [8]. In the LES framework, the nonlinear flux correction plays the role of implicit turbulence closure following the MILES approach of Grinstein and Fureby [26, ch.5], which was discussed in the ocean modeling context in [10].

A detailed description of the CABARET code on a mixed unstructured grid will be the subject of a future publication; an outline of the method on a structured Cartesian grid is

[1] Ranked #1 in temperature and #6 in velocity prediction out of 29 submissions [1]

given below. Let us consider the governing equations written in the standard conservation form:

$$\frac{\partial \mathbf{U}}{\partial t} + \frac{\partial \mathbf{F}}{\partial x} + \frac{\partial \mathbf{G}}{\partial y} + \frac{\partial \mathbf{W}}{\partial z} = \mathbf{Q}_v, \tag{1}$$

where the sources in the right-hand side include viscous terms. By mapping the physical domain to the grid space-time coordinates $(x,y,z,t) \rightarrow (i,j,k,n)$ and referring control volumes to the cell centers (fractional indices) and fluxes to the cell faces (integer indices), the algorithm proceeds from the known solution at time level (n) to the next timestep (n+1) as follows:

- Conservation predictor step:

$$\frac{\mathbf{U}^{n+\frac{1}{2}}_{i+\frac{1}{2},j+\frac{1}{2},k+\frac{1}{2}} - \mathbf{U}^{n}_{i+\frac{1}{2},j+\frac{1}{2},k+\frac{1}{2}}}{0.5\,t} + \frac{\mathbf{F}^n_{i+1} - \mathbf{F}^n_i}{\Delta x} + \frac{\mathbf{G}^n_{j+1} - \mathbf{G}^n_j}{\Delta y} + \frac{\mathbf{W}^n_{k+1} - \mathbf{W}^n_k}{\Delta z} = \mathbf{Q}^n_v \tag{2}$$

- Upwind extrapolation based on the characteristic decomposition:
  - For each cell face, decompose the conservation and flux variables into local Riemann fields, $\mathbf{U} \rightarrow R_q$, $q = 1, \ldots, N$, that correspond to the local cell-face-normal coordinate basis, where N is the dimension of the system.
  - For each cell face at the new timestep, compute a dual set of preliminary local Riemann variables that correspond to the upwind and downwind extrapolation of the characteristic fields, for example, $\tilde{R}^{n+1}_q = \left(2\,R^{n+\frac{1}{2}}_q\right)_{\text{upwind/downwind cell}} - \left(R^n_q\right)_{\text{local face}}$ for the upwind/downwind conservation volumes.
  - Correct the characteristic flux fields if they lie outside the monotonicity bounds $\left(R^{n+1}_q\right) = \max(R_q)^n$, $\tilde{R}^{n+1}_q > \max(R_q)^n$; $\left(R^{n+1}_q\right) = \min(R_q)^n$, $\tilde{R}^{n+1}_q < \min(R_q)^n$; else $\left(R^{n+1}_q\right) = \tilde{R}^{n+1}_q$.

    For reconstructing a single set of flux variables at the cell face, use an approximate Riemann solver.
- Conservation corrector step:

$$\frac{\mathbf{U}^{n+1}_{i+\frac{1}{2},j+\frac{1}{2},k+\frac{1}{2}} - \mathbf{U}^{n+\frac{1}{2}}_{i+\frac{1}{2},j+\frac{1}{2},k+\frac{1}{2}}}{0.5\,t} + \frac{\mathbf{F}^{n+1}_{i+1} - \mathbf{F}^{n+1}_i}{\Delta x} + \frac{\mathbf{G}^{n+1}_{j+1} - \mathbf{G}^{n+1}_j}{\Delta y} + \frac{\mathbf{W}^{n+1}_{k+1} - \mathbf{W}^{n+1}_k}{\Delta z} = 2\mathbf{Q}^{n+\frac{1}{2}}_v - \mathbf{Q}^n_v, \tag{3}$$

  where a second-order central approximation is used for the viscous term.

For the T-junction problem with the CABARET method, two hybrid computational grids are used with 0.53 and 4 million cells. Figure 3 shows the layout of (a) the computational domain, (b,d) the hexahedral grid with a uniform Cartesian block in the pipe center, and (c,e) a small collar area of the pipe junction covered by the mixed hexahedral/tetrahedral elements for the smaller and larger grid, respectively.

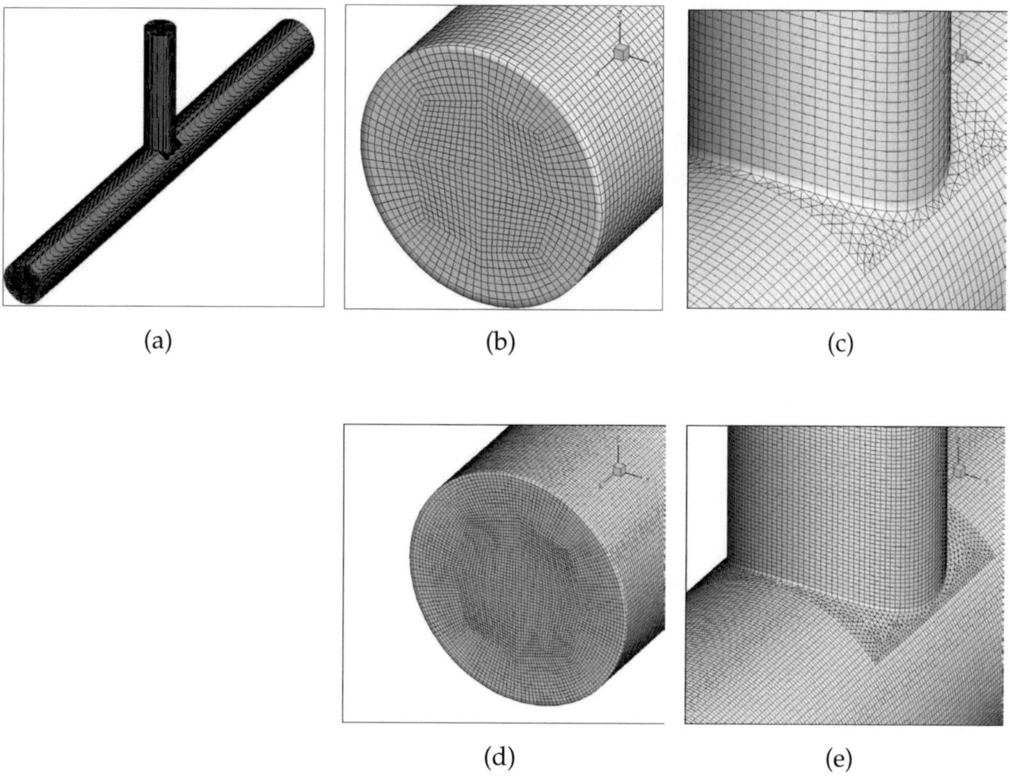

Figure 3. Computational grid used with the CABARET method: full domaim (a), pipe inlet (b,d), and mixed-grid elements in the vicinity of the junction (c,e) for the 0.5 and 4 million cell grid, respectively.

For specifying inlet boundary conditions, a recycling technique is used similar to that for Nek5000. The outflow boundary is prescribed by using characteristic boundary conditions. For the inlet boundaries at the main and the hot branch, laminar inflow conditions are specified based on prescribing the mean flow velocity profiles. The length of the pipe upstream of the junction was sufficiently far from the junction to permit an adequate turbulent flow upstream of the junction. The outflow boundary is imposed at 20 jet diameters downstream of the junction, where characteristic boundary conditions are set.

In order to speed statistical convergence, the LES CABARET solution was started from a precursor RANS k-epsilon calculation. The CABARET simulation was then run for 10 seconds to allow the statistics to settle down. This was followed by the production run during which the required solution fields were stored for a duration of 5 and 10 seconds. The computations on grids with 0.53 and 4 million cells required approximately 1.8×10^4 and 2.9×10^5 CPU hours on Lomonosov (Intel Xeon X5570 / X5670 processors).

5. Conv3D simulations

Researchers at IBRAE have been developing a 3D, unified, numerical thermal-hydraulic technique for safety analysis of the nuclear power plants, which includes (1) methods, algorithms, and software for automatically generating computing grids with local refinement near body borders; (2) methods and algorithms for solving heat and mass transfer compressible/incompressible problems for research of 3D thermal-hydraulic phenomena; and (3) approaches for modeling turbulence.

For grid construction at IBRAE an automatic technology using CAD systems for designing the computational domain has been developed. A generation of structured orthogonal/Cartesian grids with a local refinement near boundaries is incorporated into a specially developed program that has a user-friendly interface and can be utilized on parallel computers [27]. The computational technique is based on the developed algorithms with small-scheme diffusion, for which discrete approximations are constructed with the use of finite-volume methods and fully staggered grids. For modeling 3D turbulent single-phase flows, the LES approach (commutative filters) and a quasi-direct numerical simulation (QDNS) approach are used. For simulation of 3D turbulent two-phase flows by means of DNS, detailed grids and effective numerical methods developed at IBRAE for solving CFD problems are applied. For observing the interface of two-phase flow, a modified level set (LS) method and multidimensional advection/convection schemas of total variation diminishing (TVD) type with small scheme diffusion involving subgrid simulation (with local resolution) are used. The Conv3D code is fully parallelized and highly effective on high-performance computers. The developed modules were validated on a series of well-known tests with Rayleigh numbers ranging from 10^6 to 10^{16} and Reynolds numbers ranging from 10^3 to 10^5.

In order to simulate thermal-hydraulic phenomena in incompressible media, the time-dependent incompressible Navier-Stokes equations in the primitive variables [13] coupled with the energy equation are used:

$$\frac{d\vec{v}}{dt} = -\text{grad}\, p + \text{div}\, \frac{\mu}{\rho}\, \text{grad}\, \vec{v} + g, \tag{4}$$

$$\text{div}\, \vec{v} = 0 \tag{5}$$

$$\frac{\partial h}{\partial t} + \text{div}(\vec{v}\, h) = \text{div}\left(\frac{k}{\rho}\, \text{grad}\, T\right), \tag{6}$$

$$h = \int_0^T c(\xi) d\xi, \tag{7}$$

where p is pressure, normalized by the density. The basic features of the numerical algorithm [14, 28] are the following. An operator-splitting scheme for the Navier-Stokes equations is used as the predictor-corrector procedure with correction for the pressure δp:

$$\frac{v^{n+1/2} - v^n}{\tau} + \left(C(v) - \text{div}\, \frac{\mu}{\rho}\, \text{grad}\right) v^{n+1/2} + \text{grad}\, p^n - g, \tag{8}$$

$$\mathrm{div}_h \left(\frac{1}{\rho} \,\mathrm{grad}_h \, \delta p^{n+1} \right) = \frac{1}{\tau} \mathrm{div}_h \, v^{n+1/2}, \tag{9}$$

$$v^{n+1} = v^{n+1/2} - \frac{\tau}{\rho} \,\mathrm{grad}_h \, \delta p^{n+1}. \tag{10}$$

In order to construct the time-integration scheme for the energy equation, its operators are decomposed into two parts associated with the enthalpy and temperature, respectively; the following two-step procedure results:

$$\frac{h^{n+1/2} - h^n}{\tau} + \tilde{C}(u^n) \, h^{n+1/2} = 0, \tag{11}$$

$$\frac{h^{n+1} - h^{n+1/2}}{\tau} - \tilde{N} \, T^{n+1} = 0. \tag{12}$$

In the momentum equation, the operators are also split into two parts. The first part is associated with the velocity transport by convection/diffusion written in linearized form as $A_1 = C(u^n) + N$, where $N = \mathrm{div} \left(\frac{\mu}{\rho} \,\mathrm{grad}\, v \right)$. The second part deals with the pressure gradient $A_2 = \mathrm{grad}$. We note that the gradient and divergence operators are adjoints of each other (i.e., $A_2^* = -\mathrm{div}$). The additive scheme of splitting looks like the following:

$$\frac{v^{n+1/2} - v^n}{\tau} + A_1 v^{n+1/2} + A_2 p^n = f^n, \tag{13}$$

$$\frac{v^{n+1} - v^n}{\tau} + A_1 v^{n+1/2} + A_2 p^{n+1} = f^n, \tag{14}$$

$$A_2^* \, v^{n+1} = 0, \tag{15}$$

where f^n is the right-hand side. This numerical scheme is used as the predictor-corrector procedure. That is, introducing the pressure correction in Equations 14–15 leads to the well-known Poisson equation and the equation for velocity correction in the form of Equations 9–10.

In computational mathematics two variants of fictitious domain methods are recognized: continuation of coefficients at lower-order derivatives and continuation of coefficients at the highest-order derivatives. Both approaches are commonly used in computational fluid dynamics involving phase change processes. Here the first variant is employed, which in a physical sense can be considered as inclusion into the momentum equations of the model of a porous medium:

$$\frac{\partial v_\epsilon}{\partial \tau} + \tilde{N}(v_\epsilon) \, v_\epsilon - \mathrm{div} \left(\frac{\mu}{\rho} \,\mathrm{grad}\, v_\epsilon \right) + \mathrm{grad}\, p + c_\epsilon v_\epsilon = f_\epsilon, \tag{16}$$

$$\mathrm{div}\, v_\epsilon = 0. \tag{17}$$

Various formulae of c_ϵ can be employed for the flow resistance term in these equations. For Equations 16–17, the modified predictor-corrector procedure taking into account the fictitious domain method looks like the following:

$$\frac{v_\epsilon^{n+1/2} - v_\epsilon^n}{\tau} + A_1 v_\epsilon^{n+1/2} + A_2 p_\epsilon^n + c_\epsilon v_\epsilon^{n+1/2} = f^n, \tag{18}$$

$$\mathrm{div}_h \left(\frac{1}{\rho}\, \mathrm{grad}_h\, \delta p^{s+1} \right) = \mathrm{div}_h \left(\frac{1}{\rho} \frac{\tau c_\epsilon}{1 + \tau c_\epsilon} \mathrm{grad}_h\, \delta p^s \right) + \frac{1}{\tau} \mathrm{div}_h\, v_\epsilon^{n+1/2}, \tag{19}$$

$$v_\epsilon^{n+1} = v_\epsilon^{n+1/2} - \frac{1}{\rho} \frac{1}{1 + \tau c_\epsilon}\, \mathrm{grad}_h\, \delta p_\epsilon. \tag{20}$$

An iterative method with a Chebyshev set of parameters using a fast Fourier transform (FFT) solver for the Laplace operator as a preconditioner can serve as an alternative to the conjugate gradient method. The application of this approach for solving the elliptical equations with variable coefficients allows one to reach 50 times the acceleration of the commonly used method of conjugate gradients. For solving the convection problem, a regularized nonlinear monotonic operator-splitting scheme was developed [15]. A special treatment of approximation of convection terms $C(v)$ results in the discrete convective operator, which is skew-symmetric and does not contribute to the kinetic energy (i.e., is energetically neutral [15]). The numerical scheme is second, and first-order accurate in space and time, correspondingly. The algorithm is stable at a large-enough integration timestep. Details of the validation for the approach on a wide set of both 2D and 3D tests are reported in [29].

In the next section, we present the numerical results computed with Conv3D on a uniform mesh with 40 million nodes. For the sensitivity study (Section 6.2), we provide results from computation on a uniform mesh with 12 million nodes and on a nonuniform mesh with 3 and 40 million nodes with near-wall refinement. The computations on a uniform mesh with 40 million nodes required approximately 4.3×10^4 CPU hours on Lomonosov (Intel Xeon X5570 / X5670 processors).

6. Results

We focus here on a comparison of the experimental data with the numerical results from three simulation codes: Nek5000, CABARET, and Conv3D. In Section 6.2 we study the velocity field sensitivity to the computational mesh and integration time interval. A similar sensitivity study is undertaken in Section 6.3, where we investigate the effects of grid resolution and time integration interval on the reattachment region immediately downstream of the T-junction. A comparison of velocity and temperature spectra for these codes will be reported elsewhere [30].

All results presented here are nondimensionalized with the cold inlet parameters according to Table 1. For reference, the mean and rms quantities for a set of temporal values $u_n =$

$u(t = t_n)$, $n = 1, \ldots, M$, are defined as usual:

$$u = \frac{1}{M}\sum_{n=1}^{M} u_n, \qquad u' = \sqrt{\frac{1}{M}\sum_{n=1}^{M}(u_n - u)^2}\,. \tag{21}$$

6.1. Comparison with experiment

In this section, we compare the results of the three codes with experimental data. First we look at the inlet profiles in the cold and hot branches of the T-junction and then at the vertical and horizontal profiles of the mean and rms axial velocity downstream of the junction.

6.1.1. Cold and Hot Inlets

Figure 4 shows the profiles of the mean (top) and rms (bottom) streamwise velocity in the cold branch at $x = -3.1$ (left) and in the hot inlet branch at $z = 2.14286$ (i.e., $z^*/D_H = 3$) (right). The experimental data are plotted with symbols, the blue solid line represents the Nek5000 simulation, and red dashed line are the inlet profiles from the CABARET simulation. Note that the streamwise component of velocity for cold inlet coincides with the x-direction, while the hot flow in the hot inlet is in the direction opposite the z coordinate; hence, $-w$ is plotted for the hot inlet. Also note that experimental data is plotted in the same style as in [2, Figures 5 and 6 or Figures A5 and A8].

We have observed that the normalized experimental data for the cold inlet does not integrate to unity, indicating a discrepancy between the reported flow rate and the LDA measurements of approximately 6%. On the contrary, using the trapezoidal rule, the integrals of Nek5000 profiles for the cold and hot inlets are equal to 1.0027 and 1.3080, respectively, once averaged over and normalized by the corresponding inlet cross-sections. The same integration procedure for CABARET profiles in Figure 4 gives the normalized inlet flow rates equal to 1.014 and 1.330 for the cold and hot inlets, respectively. These values are in a good agreement with the nondimensional values for inlet velocities $U = 1.000$ and $U_H = 1.307$ given in Table 2. Note that the normalized inlet mass flow rates for Nek5000 were averaged over the $\xi = z$ line ($y = 0$) and $\xi = y$ line ($z = 0$) for the cold inlet and over the $\xi = x$ line ($y = 0$) and $\xi = y$ line ($x = 0$) for the hot inlet (Figure 4).

Moreover, a comparison of the shape of the inlet simulation profiles with the experimental data reveals a few differences. The shape of the hot inlet profile for the mean and rms velocity for Nek5000 simulation is not as flat as the experimental or CABARET profiles. This difference can be attributed to a particular modeling of the non-fully-developed flow with the recycling technique described in Section 3. Note that the same technique works nearly perfectly for the cold inlet, where the profile of the mean velocity for Nek5000 is in excellent agreement with the experimental data points after multiplication by 0.94 factor to account for the mass conservation uncertainty of 6%. Similarly, the cold inlet rms velocity for Nek5000 data shows good agreement, diverging from the experimental data points only in the near-wall region, which can be explained by the lower Reynolds number of the simulation, $Re = 4 \times 10^5$ (cf. Table 2). These results indicate that further investigation is needed into the effects of the non-fully-developed flow in the hot inlet for Nek5000 simulation.

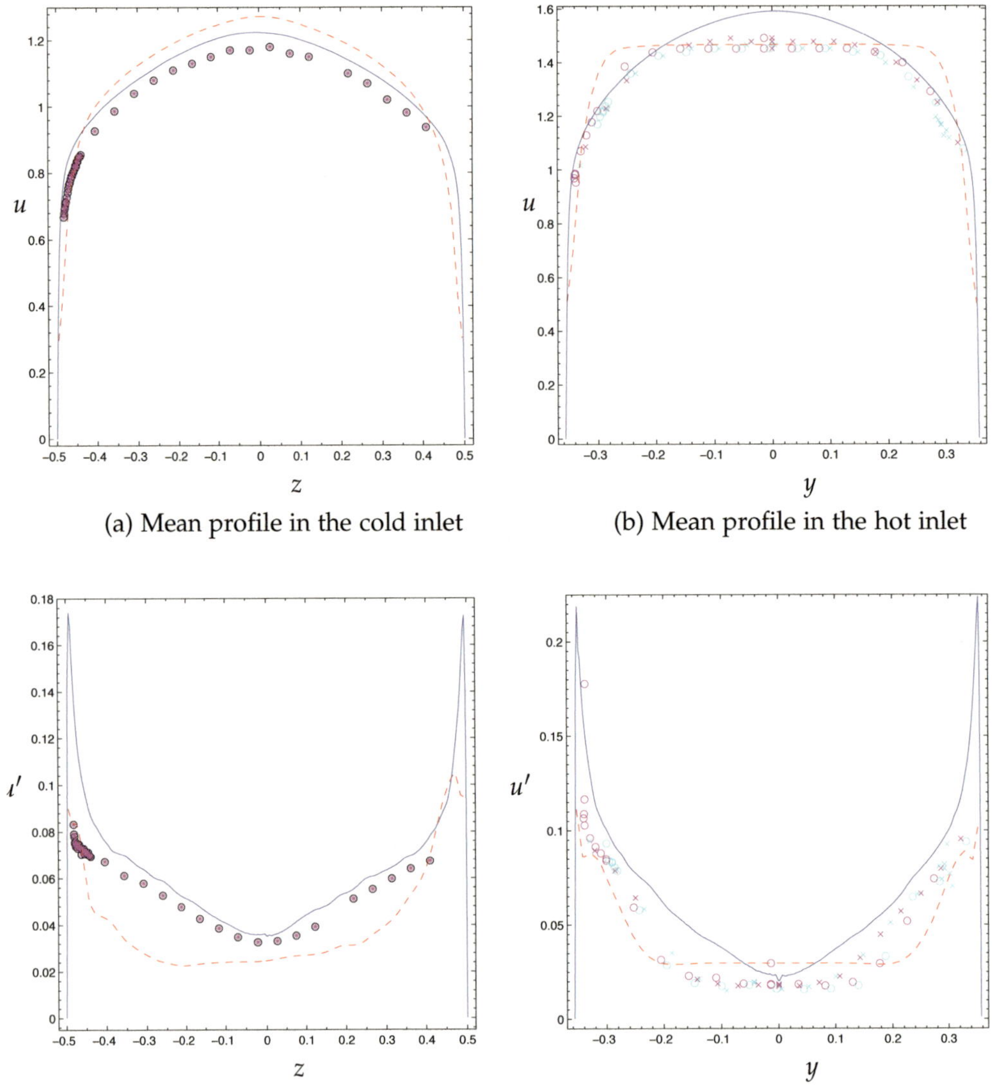

(a) Mean profile in the cold inlet (b) Mean profile in the hot inlet

(c) Rms profile in the cold inlet (d) Rms profile in the hot inlet

Figure 4. Mean and rms velocity profiles in the cold inlet branch and in the hot inlet branch vs a centerline coordinate ξ for the experimental data (symbols) and simulation results with Nek5000 (blue solid line) and with CABARET (red dashed line).

On the contrary, the CABARET simulation models the flat profile at the hot inlet well (Figure 4, right). The cold inlet profiles, however, deviate substantially from the experimental data, especially in the case of rms profiles (Figure 4, left).

6.1.2. Downstream of the T-Junction

We next look at the axial mean u and rms u' velocity downstream of the T-junction. Figure 5 shows a "bird's-eye" view of the mean (u) and rms (u') velocity profiles downstream of the T-junction at x=0.6, 1.6, 2.6, 3.6, and 4.6. Here the experimental data

http://dx.doi.org/10.5772/53143

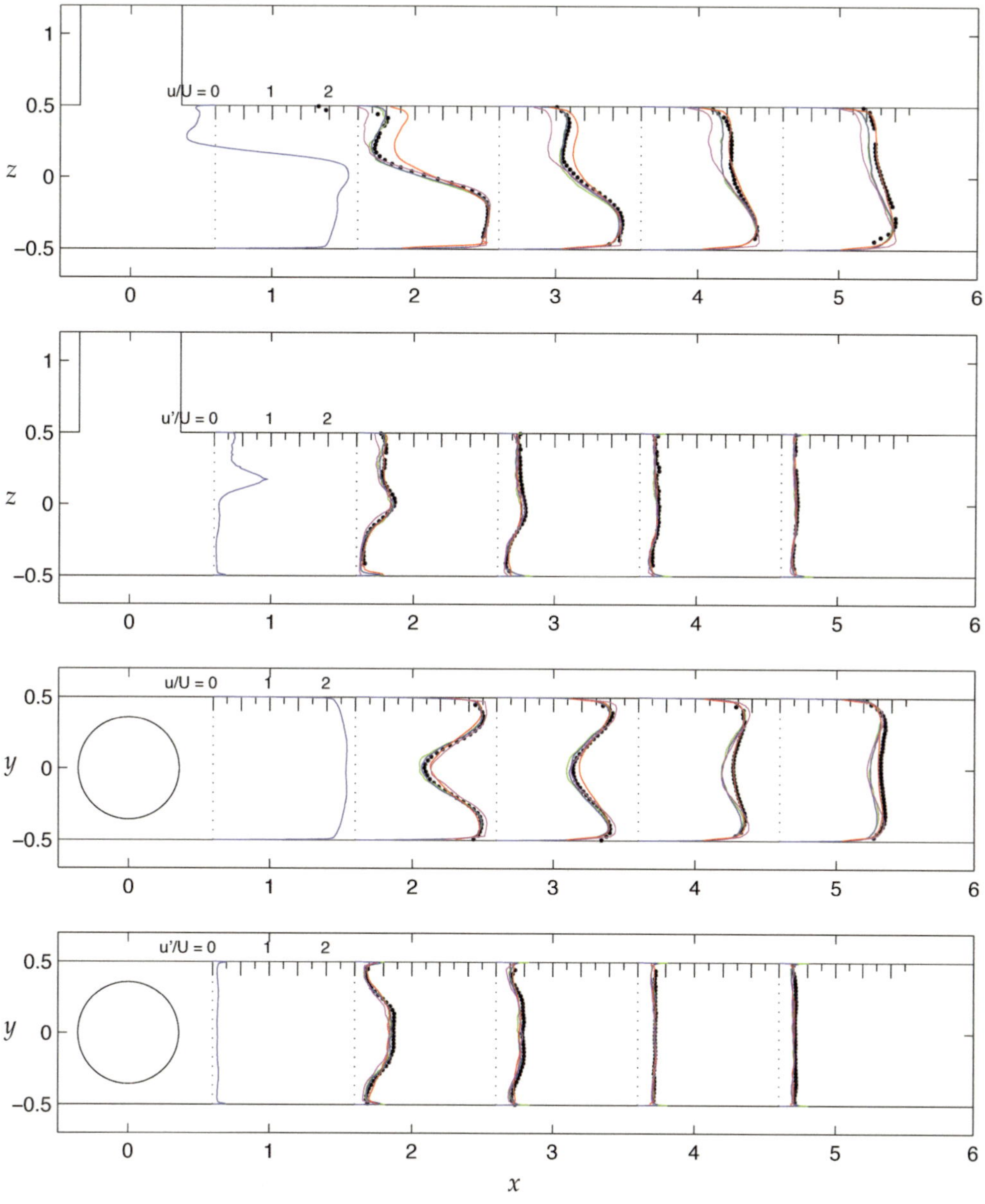

Figure 5. T-junction experimental data (·), CABARET (red), and Conv3D (magenta) results, and Nek5000 simulation for the benchmark submission results (green) and for the calculation with a longer time integration (blue line). From top to bottom: vertical profiles of axial mean (u) and rms (u') velocity and their horizontal profiles.

(black symbols) are contrasted with numerical simulations with CABARET (red), Conv3D (magenta), and Nek5000 for the benchmark submission results (green) and for longer time integration/averaging (blue). Note the T-junction geometry outline and the equal unit scale for the mean and rms axial velocity equal to the velocity scale, namely, the axial mean velocity in the cold inlet branch $U = 1.000$ (Table 2). For reference, we provide a detailed

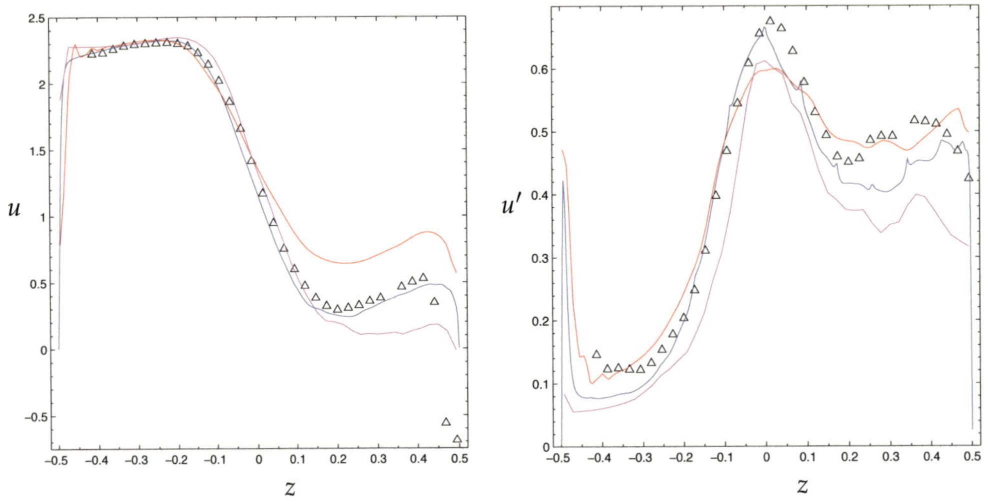

(a) Mean vertical profile at $x = 1.6$ (b) Rms vertical profile at $x = 1.6$

Figure 6. Axial mean velocity and rms vertical profiles at x=1.6 for experimental data (triangles) and simulation with Nek5000 (blue), CABARET (red), and Conv3D (magenta line).

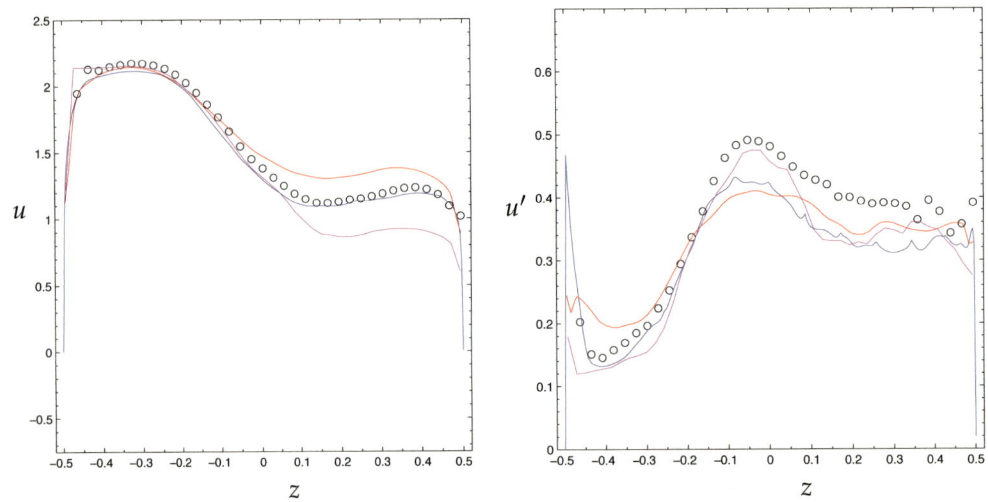

(a) Mean vertical profile at $x = 2.6$ (b) Rms vertical profile at $x = 2.6$

Figure 7. Axial mean velocity and rms vertical profiles at x=2.6 for experimental data (circles) and simulation with Nek5000 (blue), CABARET (red), and Conv3D (magenta line).

comparison of numerical simulations with experimental data points of the mean and rms axial velocity for each cross-section separately (Figures 6–13). Each figure shows the results of numerical simulations with Nek5000 (blue), CABARET (red), and Conv3D (magenta) against the experimental data points (symbols). The vertical profiles of the mean u (left) and rms u' (right) axial velocity are shown in Figures 6–9 at $x = 1.6 \ldots 4.6$, correspondingly, while the

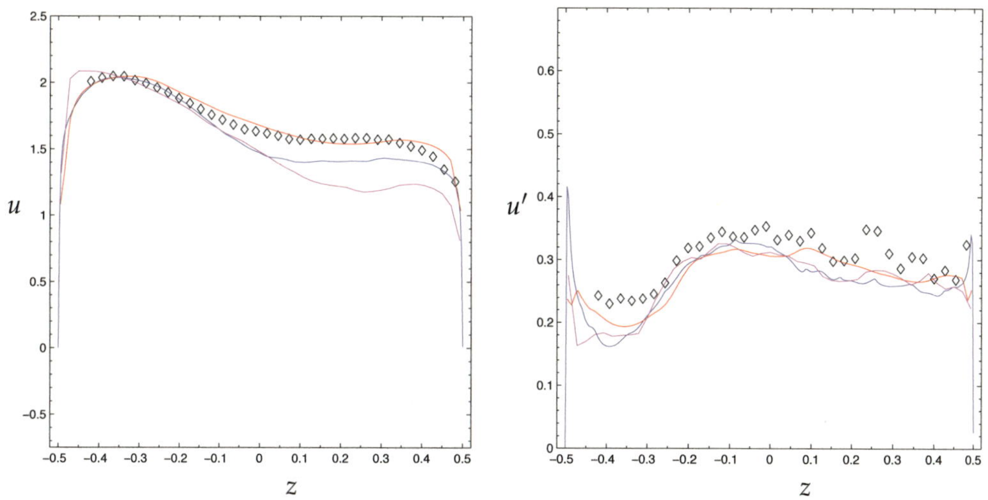

(a) Mean vertical profile at $x = 3.6$

(b) Rms vertical profile at $x = 3.6$

Figure 8. Axial mean velocity and rms vertical profiles at x=3.6 for experimental data (diamonds) and simulation with Nek5000 (blue), CABARET (red), and Conv3D (magenta line).

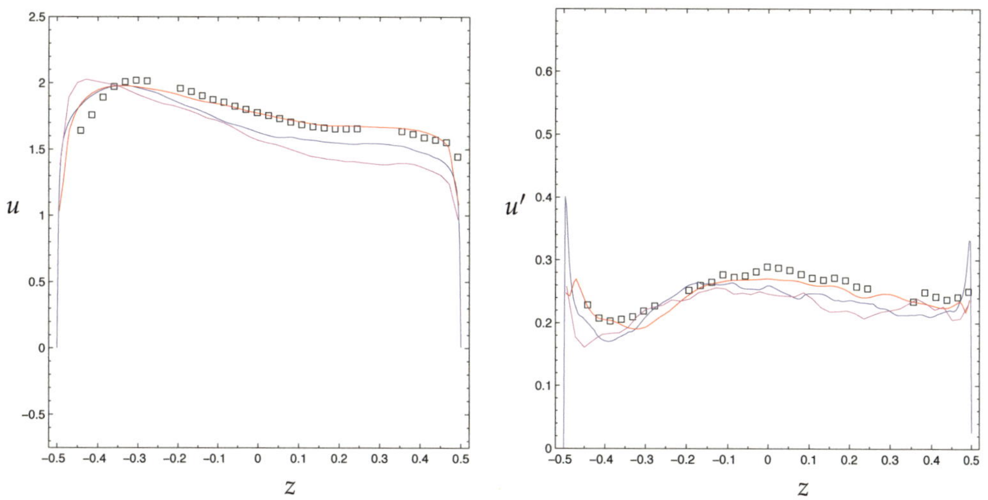

(a) Mean vertical profile at $x = 4.6$

(b) Rms vertical profile at $x = 4.6$

Figure 9. Axial mean velocity and rms vertical profiles at x=4.6 for experimental data (squares) and simulation with Nek5000 (blue), CABARET (red), and Conv3D (magenta line).

horizontal profiles are plotted similarly in Figures 10–13. Detailed comparison with Nek5000 benchmark submission results is reported in Section 6.2 (see Figure 14).

It is encouraging that overall agreement of simulation results for the three codes with the experimential data is good for the both mean and rms axial velocity. The Nek5000 profiles (blue lines) of the rms velocity match experimental data points well (Figures 5 and 6–13,

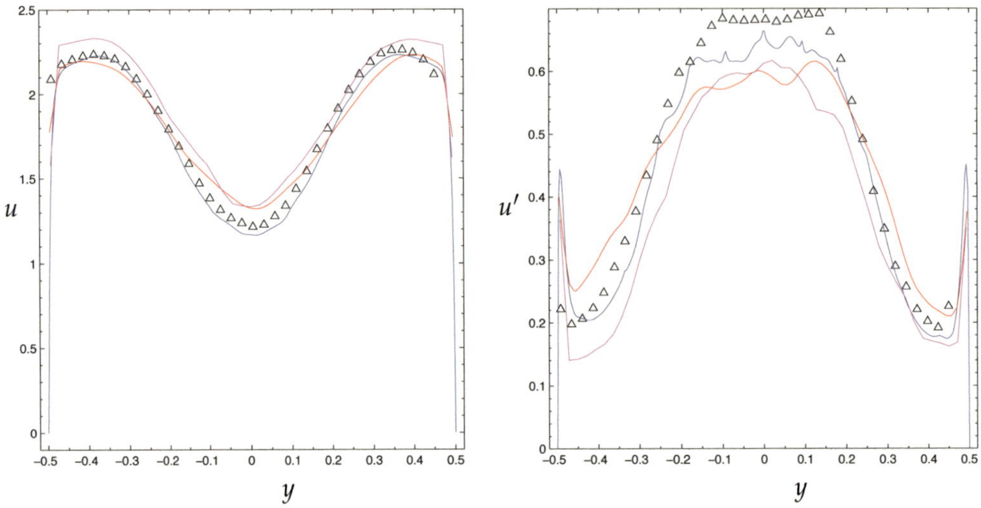

(a) Mean horizontal profile at $x = 1.6$ (b) Rms horizontal profile at $x = 1.6$

Figure 10. Axial mean velocity and rms horizontal profiles at x=1.6 for experimental data (triangles) and simulation with Nek5000 (blue), CABARET (red), and Conv3D (magenta line).

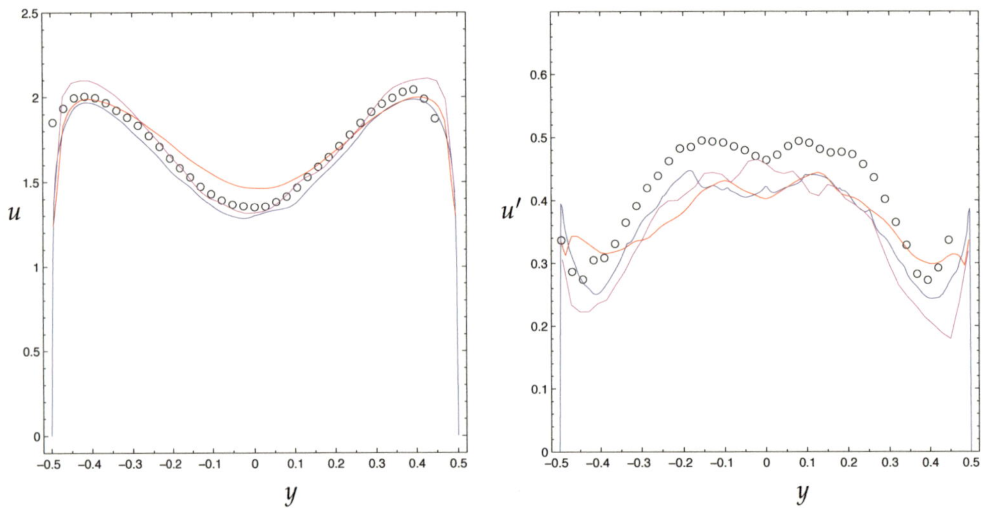

(a) Mean horizontal profile at $x = 2.6$ (b) Rms horizontal profile at $x = 2.6$

Figure 11. Axial mean velocity and rms horizontal profiles at x=2.6 for experimental data (circles) and simulation with Nek5000 (blue), CABARET (red), and Conv3D (magenta line).

right). Moreover, the agreement between the simulation and experiment for the mean velocity (left figures) is best at $x = 2.6$ (Figures 7 and 11) and at $x = 1.6$ (Figures 6 and 10) apart of two near-wall data points close to at $z = 0.5$ (Figure 6). This discrepancy is the focus of an investigation described in Section 6.3. The deviation of the mean velocity for Nek5000 results from experimental data further downstream at $x = 3.6$ and 4.6 in Figures 5, 8–9, and

http://dx.doi.org/10.5772/53143

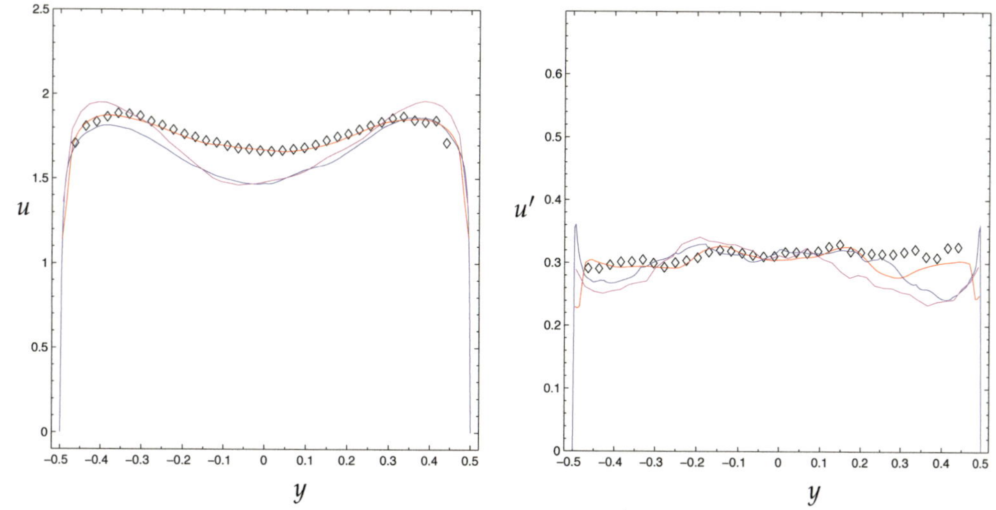

(a) Mean horizontal profile at $x = 3.6$ (b) Rms horizontal profile at $x = 3.6$

Figure 12. Axial mean velocity and rms horizontal profiles at x=3.6 for experimental data (diamonds) and simulation with Nek5000 (blue), CABARET (red), and Conv3D (magenta line).

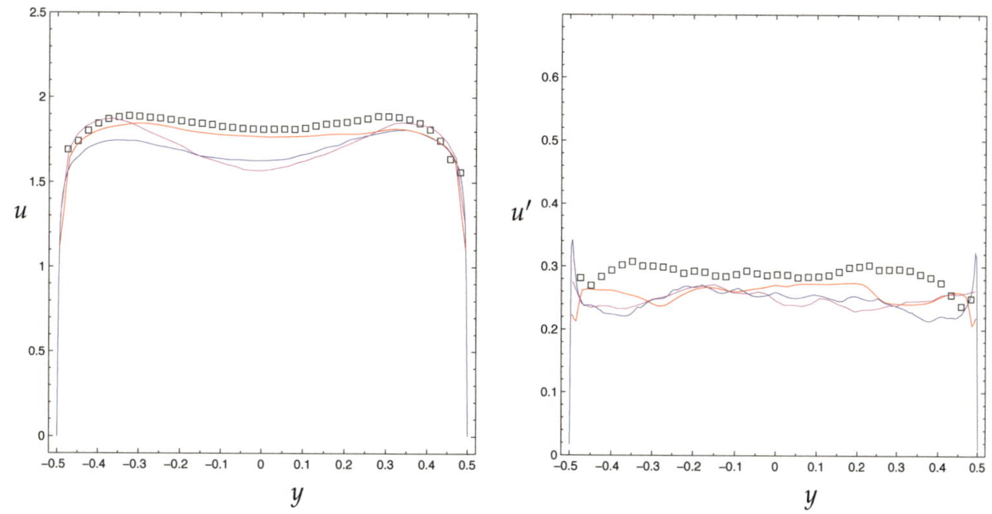

(a) Mean horizontal profile at $x = 3.6$ (b) Rms horizontal profile at $x = 3.6$

Figure 13. Axial mean velocity and rms horizontal profiles at x=4.6 for experimental data (squares) and simulation with Nek5000 (blue), CABARET (red), and Conv3D (magenta line).

12–13 can be attributed to the lower Reynolds number used in the simulation because of the time constraints and will be the subject of a follow-up study.

On the contrary, CABARET simulation results (red lines) agree with experiment remarkably well further downstream, at $x = 3.6$ (Figures 8 and 12) and $x = 4.6$ (Figures 9 and 13), with the notable exception of the best match of rms data with experimental points in the recirculation region at $x = 1.6$ and $z > 0$ (Figure 6). However, close to the T-junction, at x=1.6 and 2.6, the CABARET profiles of the mean axial velocity deviate from experimental points at $z > 0$ (Figures 6–7) and near the centerline $y = 0$ (Figures 10–11). Similar to CABARET profiles, the Conv3D results (magenta lines) show the most deviation from experimental data for the mean axial velocity at $z > 0$ (Figures 6–9) and near the centerline $y = 0$ (Figures 10–13). However, the agreement of Conv3D simulation with experiment is best at $x = 4.6$ and $0.4 < |y| < 0.5$ (Figure 13).

6.2. Sensitivity study

To study the effects of increasing resolution and time averaging interval, we performed an additional set of simulations with Nek5000, CABARET, and Conv3D.

The Figures 14–16 highlight the mesh sensitivity of velocity profiles extracted from results of numerical simulations with Nek5000, CABARET and Conv3D, correspondingly. Each figure compares the vertical and horizontal profiles of the mean axial velocity u and its rms u' for different meshes with the experimental data. More detailed comparisons can be found in [31].

In addition to the benchmark submission results with $N = 7$ (i.e., $n \approx EN^3 \approx 2.1 \times 10^7$ points), Nek5000 runs were conducted with $N = 5$ ($n \approx 7.7 \times 10^6$ points). This case started with the $N = 7$ results and was run with $\Delta t = 5 \times 10^{-4}$ and averaged over about 110 convective time units (i.e., about 26 seconds). Figure 14 shows the vertical (left) and horizontal (right) profiles of the mean (top) and rms (bottom) axial velocity profiles at x=0.6 (magenta), 1.6 (black), 2.6 (blue), 3.6 (green), and 4.6 (red) with $N = 5$ (dashed) and $N = 7$ (solid). The benchmark submission results (i.e., with N=7 and shorter time average) plotted with dash-dotted line at $x = 1.6 \ldots 4.6$ are in the excellent agreement with the longer time average run (solid). All profiles agree well with the experimental data points (symbols), especially considering the fact that the Reynolds number of these simulations, $Re = 4 \times 10^5$, is two times less than that of the experiment (see Table 2). In general, the profiles from the coarse mesh simulation (i.e., $N = 5$, dashed) are close to the solution for the finer mesh (solid and dash-dotted lines). Note the excellent agreement between the profiles for $N = 5$ and $N = 7$ at $x = 0.6$, where the strength of the reversed flow is close to its peak (Figure 17). The largest deviation in the profiles (up to about 0.2) is observed at $x = 1.6$ and $0.1 < z < 0.25$; thus, a further study is warranted with even finer mesh, say, $N = 8$ or $N = 9$.

Similarly, for the CABARET simulations, the vertical and horizontal profiles of u and u' are plotted in Figure 15 at $x = 1.6 \ldots 4.6$. These figures show results from CABARET calculations on a coarser mesh with 0.5 million points averaged over a half time interval (red dotted) and the full time interval (blue dash-dotted) and on a finer mesh with 4 million points averaged over a half time interval (green dashed) and the full time interval (solid black line). Note the excellent convergence for the mean axial velocity profiles at $x = 3.6$ and $x = 4.6$.

For the Conv3D simulations, the vertical and horizontal profiles of u and u' are plotted in Figure 16 at $x = 1.6 \ldots 4.6$. This figure shows results from Conv3D simulations on a 40 million node uniform (black solid) and nonuniform (blue dash-dotted) mesh, on a 12 million node uniform mesh (green dashed), and on a 3 million node nonuniform mesh (red dotted line). Note the good convergence for the mean axial velocity profiles at $x = 1.6 \ldots 3.6$ and $z < 0$.

http://dx.doi.org/10.5772/53143

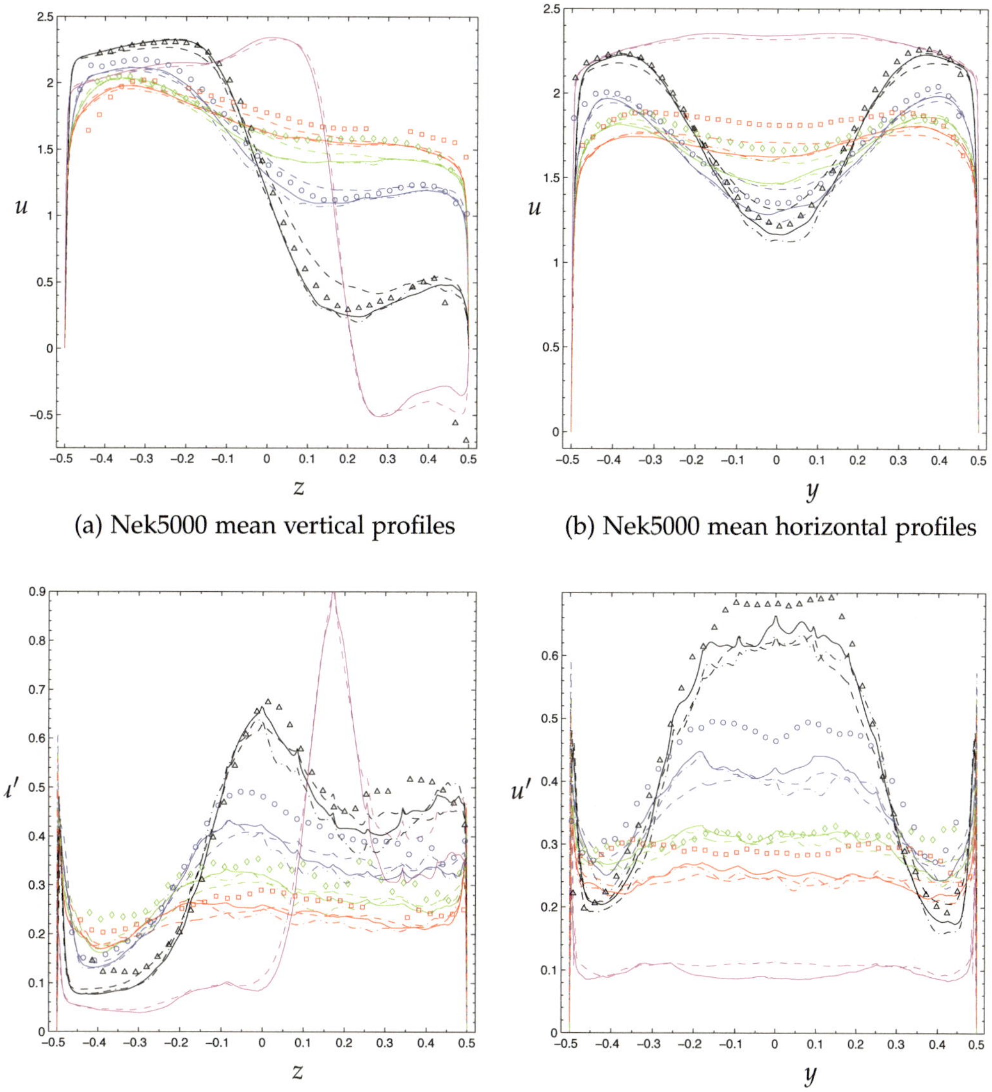

(a) Nek5000 mean vertical profiles

(b) Nek5000 mean horizontal profiles

(c) Nek5000 rms vertical profiles

(d) Nek5000 rms horizontal profiles

Figure 14. Mean and rms profiles of axial velocity at x=0.6 (magenta), 1.6 (black), 2.6 (blue), 3.6 (green), and 4.6 (red) for the experimental data (symbols) and Nek5000 simulation with $N = 5$ (dashed) and $N = 7$ (solid) and for Nek5000 benchmark submission results (dash-dotted).

6.3. Study of reversed flow region

To address the discrepancy between experimental data and simulations near the upper wall at $x = 1.6$, we have also undertaken a series of Nek5000 simulations to investigate the sensitivity of the reattachment of the recirculation region downstream of the T-junction.

Summarizing the effects of Reynolds number, grid resolution, and time averaging on the reversed flow region, Figure 17 shows time-averaged velocity profiles near the upper wall at z=0.005 and y=0 for Nek5000 simulations at $Re = 9 \times 10^4$ with $N = 11$ (red), at $Re = 6 \times 10^4$

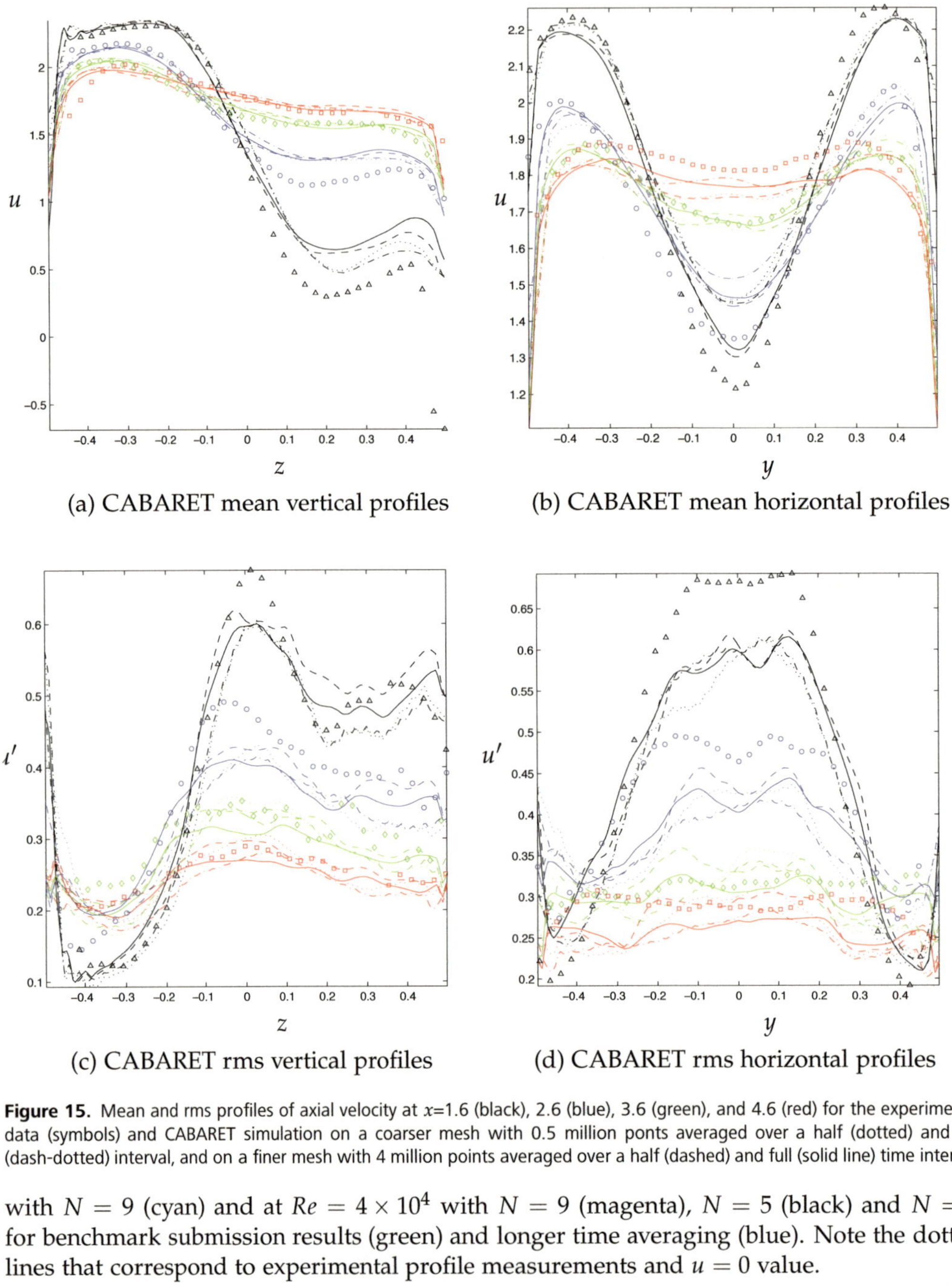

(a) CABARET mean vertical profiles

(b) CABARET mean horizontal profiles

(c) CABARET rms vertical profiles

(d) CABARET rms horizontal profiles

Figure 15. Mean and rms profiles of axial velocity at x=1.6 (black), 2.6 (blue), 3.6 (green), and 4.6 (red) for the experimental data (symbols) and CABARET simulation on a coarser mesh with 0.5 million ponts averaged over a half (dotted) and full (dash-dotted) interval, and on a finer mesh with 4 million points averaged over a half (dashed) and full (solid line) time interval.

with $N = 9$ (cyan) and at $Re = 4 \times 10^4$ with $N = 9$ (magenta), $N = 5$ (black) and $N = 7$ for benchmark submission results (green) and longer time averaging (blue). Note the dotted lines that correspond to experimental profile measurements and $u = 0$ value.

Based on this study, we conclude that the reattachment region is insensitive mainly to an increase in Reynolds number and grid resolution. Moreover, results at $Re = 4 \times 10^4$ indicate that the recirculation region is not sensitive to the averaging/integration time interval. Ideally, one should conduct a further investigation at higher Reynolds number

http://dx.doi.org/10.5772/53143

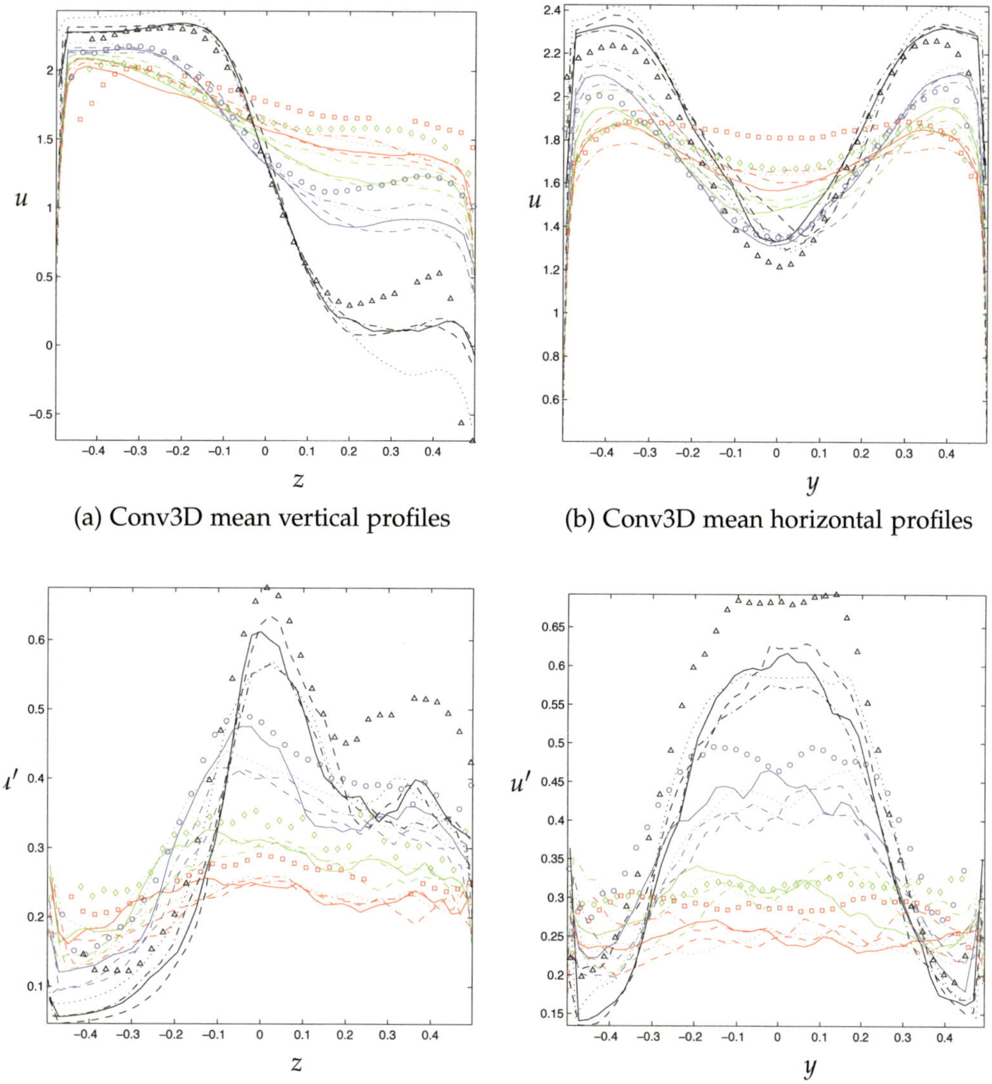

(a) Conv3D mean vertical profiles

(b) Conv3D mean horizontal profiles

(c) Conv3D rms vertical profiles

(d) Conv3D rms horizontal profiles

Figure 16. Mean and rms profiles of axial velocity at x=1.6 (black), 2.6 (blue), 3.6 (green), and 4.6 (red) for the experimental data (symbols) and Conv3D simulation on 40 milliom node uniform (solid) and nonuniform (dash-dotted) mesh, on 12 million node uniform mesh (dashed) and on 3 million node nonuniform mesh (dotted line).

with longer integration/averaging time to resolve the discrepancy between the simulations and experimental data points near the upper wall at $x = 1.6$. For now, we quote that in the experimental measurements of velocity, "...the focus ... is not the near-wall region" [2, p. 15]. This underscores a necessity of conducting concurrent experiments and validation simulations in which discrepancies like the extent of reciculation region or flow rate uncertainty arising here could be easier to detect and additional data acquisitions and simulations could be performed in order to resolve them.

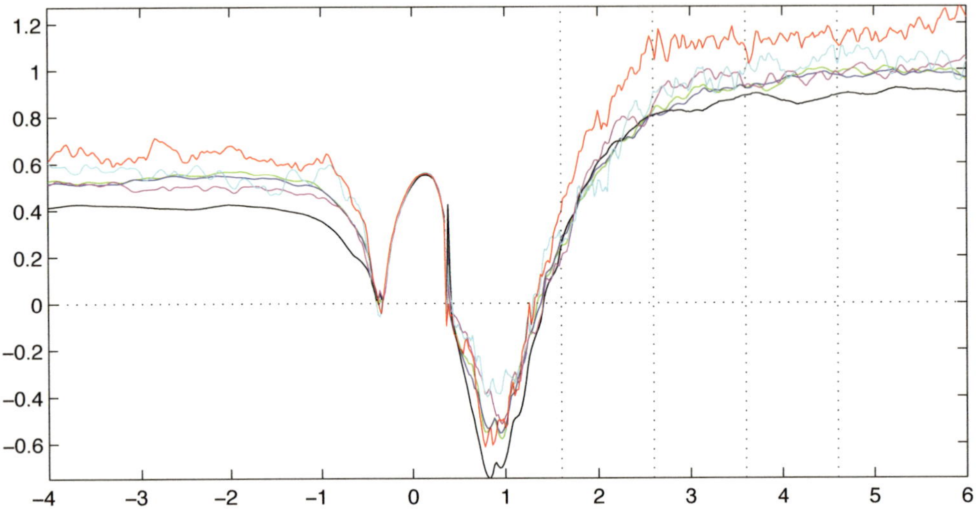

Figure 17. Axial mean velocity profile near the upper wall at z=0.0050 and y=0.0000 for Nek5000 simulation at $Re = 9 \times 10^4$ with $N = 11$ (red), at $Re = 6 \times 10^4$ with $N = 9$ (cyan) and at $Re = 4 \times 10^4$ with $N = 9$ (magenta), $N = 5$ (black) and $N = 7$ for the benchmark submission results (green) and longer time averaging (blue).

7. Conclusions and future work

Several fully unsteady computational models in the framework of large eddy simulations (LES) are implemented for a thermohydraulic transport problem relevant to the design of nuclear power plant pipe systems. Specifically, numerical simulations for the recent 2010 OECD/NEA Vattenfall T-junction benchmark problem concerned with thermal stripping in a T-junction have been conducted with three numerical techniques based on finite-difference implicit LES (CABARET code), finite-volume LES (CONV3D code) and spectral-element (Nek5000 code) approaches. The simulation results of all three methods, including the blind test submission results of Nek5000, show encouraging agreement with the experiment despite some differences in the operating conditions between the simulations and the experiment. In particular, the Nek5000 results tend to closely match the experiment data close to the T-junction, and the CABARET solution well captures the experimental profiles farther downstream of the junction. The differences in the operating conditions between the simulation and the experiment include the uncertainty of the mass flow rate reported in the experiment (about 6%), a reduction in the effective Reynolds number used in some of the simulations that was needed to speedup the computations (Nek5000), and some discrepancies in the inflow boundary conditions. Nevertheless, the good level of agreement between the current simulations and the experiment despite these discrepancies indicates that the flow dynamics in the T-junction experiment is driven mainly by large-scale mixing effects that were not very sensitive to the differences in the operating conditions.

For further investigation, the improvement of the quality of turbulent inflow conditions (most notably, at the hot inlet boundary for Nek5000 and at the cold inlet for CABARET) is planned as well as running additional high-resolution simulations to capture the high Reynolds number regimes typical of the experiment. A further study also will be devoted to a detailed analysis of the temporal spectra of velocity and temperature fluctuations obtained from all three LES solutions.

Acknowledgments

This work was partially supported by the National Institutes of Health, RO1 Research Project Grant (2RO1HL55296-04A2), by Whitaker Foundation Grant (RG-01-0198), and by the Office of Science, U.S. Department of Energy, under Contract DE-AC02-06CH11357. The work was completed as part of the SHARP reactor performance and safety code suite development project, which is funded under the auspices of the Nuclear Energy Advanced Modeling and Simulation (NEAMS) program of the U.S. Department of Energy Office of Nuclear Energy.

The authors are grateful to Gail Pieper for editing the manuscript. One of the authors also thanks Gary Leaf, Elia Merzari, and Dave Pointer for helpful discussions, Scott Parker and other staff at Argonne Leadership Computing Facility, and his wife Alla for her patience during his work on this project.

Author details

Aleksandr V. Obabko[1], Paul F. Fischer[1],
Timothy J. Tautges[1], Vasily M. Goloviznin[2],
Mikhail A. Zaytsev[2], Vladimir V. Chudanov[2], Valeriy A. Pervichko[2],
Anna E. Aksenova[2] and
Sergey A. Karabasov[3,4]

1 Mathematics and Computer Science Division, Argonne National Laboratory, Argonne, IL, USA
2 Moscow Institute of Nuclear Energy Safety (IBRAE), Moscow, Russia
3 Queen Mary University of London, School of Engineering and Materials Science, London, UK
4 Cambridge University, Cambridge, UK

References

[1] J. H. Mahaffy and B. L. Smith. Synthesis of benchmark results. In *Experimental Validation and Application of CFD and CMFD Codes to Nuclear Reactor Safety Issues (CFD4NRS-3)*, Bathesda, MD, 2010.

[2] B. L. Smith, J. H. Mahaffy, K. Angele, and J. Westin. Experiments and unsteady CFD-calculations of thermal mixing in a T-junction. *Nuclear Engineering and Design*, 2012. To appear.

[3] A.T. Patera. A spectral element method for fluid dynamics: laminar flow in a channel expansion. *J. Comput. Phys.*, 54:468–488, 1984.

[4] M.O. Deville, P.F. Fischer, and E.H. Mund. *High-order methods for incompressible fluid flow.* Cambridge University Press, Cambridge, 2002.

[5] A.A. Samarskii and V.M. Goloviznin. Difference approximation of convective transport with spatial splitting of time derivative. *Mathematical Modelling*, 10(1):86–100, 1998.

[6] A. Iserles. Generalized leapfrog methods. *IMA Journal of Numerical Analysis*, 6(3):381–392, 1986.

[7] Philip Roe. Linear bicharacteristic schemes without dissipation. *SIAM Journal on Scientific Computing*, 19(5):1405–1427, September 1998.

[8] S. A. Karabasov and V. M. Goloviznin. Compact accurately boundary-adjusting high-resolution technique for fluid dynamics. *J. Comput. Phys.*, 228:7426–7451, October 2009.

[9] S.A. Karabasov and V.M. Goloviznin. New efficient high-resolution method for non-linear problems in aeroacoustics. *AIAA Journal*, 45(12):2861–2871, 2007.

[10] S.A. Karabasov, P.S. Berloff, and V.M. Goloviznin. CABARET in the ocean gyres. *Ocean Modelling*, 30(2-3):155–168, 2009.

[11] S. A. Karabasov, V. M. Goloviznin, and M.A. Zaytsev. Towards empiricism-free large eddy simulation for thermo-hydraulic problems. In *CFD for Nuclear Reactor Safety Applications (CFD4NRS-3)*, Bathesda, MD, September 2010.

[12] G. A. Faranosov, V. M. Goloviznin, S. A. Karabasov, V. G. Kondakov, V. F. Kopiev, and M. A. Zaytsev. CABARET method on unstructured hexahedral grids for jet noise computation. In *18th AIAA/CEAS Aeroacoustics Conference*, Colorado Springs, CO, 2012.

[13] L.A. Bolshov, V.V. Chudanov, V.F. Strizhov, S.V. Alekseenko, V.G. Meledin, and N.A. Pribaturin. Best estimate methodology for modeling bubble flows. In *14th International Conference on Nuclear Engineering (ICONE14) CD-ROM*, number ICONE-89296, Miami, Florida, 2006.

[14] V.V. Chudanov, A.E. Aksenova, and V.A. Pervichko. 3D unified CFD approach to thermalhydraulic problems in safety analysis. In *IAEA Technical Meeting on Use of Computational Fluid Dynamics (CFD) Codes for Safety Analysis of Nuclear Reactor Systems, Including Containment CD-ROM*, number IAEA-TECDOC-1379, Pisa, Italy, 2003.

[15] P. N. Vabishchevich, V. A. Pervichko, A. A. Samarskii, and V. V. Chudanov. *Journal of Computing Mathematics and of Mathematical Physics*, 40:860–867, 2000.

[16] J. Westin, F. Alavyoon, L. Andersson, P. Veber, M. Henriksson, and C. Andersson. Experiments and unsteady CFD-calculations of thermal mixing in a T-junction. In *OECD/NEA/IAEA Benchmarking of CFD Codes for Application to Nuclear Reactor Safety (CFD4NRS) CD-ROM*, Munich, Germany, 2006.

[17] Y. Maday and A.T. Patera. Spectral element methods for the Navier-Stokes equations. In A.K. Noor and J.T. Oden, editors, *State-of-the-Art Surveys in Computational Mechanics*, pages 71–143. ASME, New York, 1989.

[18] P.F. Fischer and A.T. Patera. Parallel spectral element of the Stokes problem. *J. Comput. Phys.*, 92:380–421, 1991.

[19] P.F. Fischer. An overlapping Schwarz method for spectral element solution of the incompressible Navier-Stokes equations. *J. Comput. Phys.*, 133:84–101, 1997.

[20] A.G. Tomboulides, J.C.Y. Lee, and S.A. Orszag. Numerical simulation of low Mach number reactive flows. *Journal of Scientific Computing*, 12:139–167, June 1997.

[21] A.G. Tomboulides and S.A. Orszag. A quasi-two-dimensional benchmark problem for low Mach number compressible codes. *J. Comput. Phys.*, 146:691–706, 1998.

[22] A.G. Tomboulides, M. Israeli, and G.E. Karniadakis. Efficient removal of boundary-divergence errors in time-splitting methods. *J. Sci. Comput.*, 4:291–308, 1989.

[23] T. J. Tautges, P. F. Fischer, I. Grindeanu, R. Jain, V. Mahadevan, A. Obabko, M. A. Smith, S. Hamilton, K. Clarno, M. Baird, and M. Berrill. A coupled thermal/hydraulics - neutronics - fuel performance analysis of an SFR fuel assembly. Technical report, ANL/NE-12/53, Argonne National Laboratory, 2012.

[24] V.M. Goloviznin. Balanced characteristic method for systems of hyperbolic conservation laws. *Doklady Mathematics*, 72(1):619–623, 2005.

[25] V.M. Goloviznin, V.N. Semenov, I.A Korortkin, and S.A. Karabasov. A novel computational method for modelling stochastic advection in heterogeneous media. *Transport in Porous Media*, 66(3):439–456, 2007.

[26] F.F. Grinstein, L.G. Margolin, and W.J. Rider. *Implicit Large Eddy Simulation: Computing Turbulent Fluid Dynamics*. Cambridge University Press, Cambridge, 2007.

[27] V.V. Chudanov. The integrated approach to a solution of CFD problems in complicated areas. *Izvestiya of Academy of Sciences*, 6:126–135, 1999. (in Russian).

[28] V.V. Chudanov, A.E. Aksenova, and V.A. Pervichko. CFD to modelling molten core behaviour simultaneously with chemical phenomena. In *11th International Topical Meeting on Nuclear Reactor Thermal-Hydraulics (NURETH-11) CD-ROM*, number 048, Avignon, France, 2005.

[29] V.V. Chudanov, A.E. Aksenova, and V.A. Pervichko. Methods of direct numerical simulation of turbulence in thermalhydraulic's problems of fuel assembly. *Izvestiya Rossiiskoi Akademii Nauk. Seriya Energetica*, 6:47–57, 2007.

[30] A. V. Obabko, P. F. Fischer, S. Karabasov, V. M. Goloviznin, M. A. Zaytsev, V. V. Chudanov, V. A. Pervichko, and A. E. Aksenova. LES spectrum comparison study for OECD/NEA T-junction benchmark. Technical Report ANL/MCS-TM-327, Argonne National Laboratory, 2012.

[31] A. V. Obabko, P. F. Fischer, T. J. Tautges, S. Karabasov, V. M. Goloviznin, M. A. Zaytsev, V. V. Chudanov, V. A. Pervichko, and A. E. Aksenova. CFD validation in OECD/NEA T-junction benchmark. Technical Report ANL/NE-11/25, Argonne National Laboratory, 2011.

Chapter 3

CFD as a Tool for the Analysis of the Mechanical Integrity of Light Water Nuclear Reactors

Hernan Tinoco

Additional information is available at the end of the chapter

http://dx.doi.org/10.5772/52691

1. Introduction

The analysis of the mechanical integrity of Light Water Nuclear Reactors has experienced a strong evolution in the two latest decades. Until the nineteen eighties, the structural design was practically based on static loads with amplifying factors to take dynamic effects into account, (see e. g. Laheyand Moody, 1993), and on wide safety margins. Even if an incipient methodology for studying earthquake effects already existed, it was only at the end of the nineties when a complete analysis about dynamic loads, caused by a steam line guillotine-break of a BWR, was carried out for the first time (Hermansson and Thorsson, 1997). This analysis, based on a three-dimensional Computational Fluid Dynamic (CFD) simulation of the break, revealed a discrepancy with former applied loads that had never been exposed in the past due to inappropriate experimental setup geometry (Tinoco, 2001). In addition, the analysis pointed out the importance of also studying the load frequencies. The complete issue is further discussed in Section 3.

Nonetheless, CFD has shown to be a much more versatile tool. Issues like checking the maximum temperature of the core shroud of the reactor, when the power of the reactor is uprated (Tinoco et al., 2008), or estimating the thermal loadings during different transient conditions when the Emergency Core Cooling System (ECCS) has been modified, (Tinoco et al., 2005), or studying the automatic boron injection during an Anticipated Transient Without Scram (ATWS),(Tinoco et al., 2010a), have all been settled by means of CFD simulations.

Even more significantly, a rather complete CFD model of the Downcomer, internal Main Recirculation Pumps (MRP) and Lower Plenum up to the Core Inlet, (Tinoco and Ahlinder, 2009), has shed some light concerning the thermal mixing in the different regions of the reactor and also about the appropriateness of using connected sub-models as constituting parts of a larger model. The complete matter of thermal mixing is discussed in detail in Section 4.

Last but not least, CFD has demonstrated that it can be used to establish the causes of component failure. Even if justified suspicions existed about the cause of two broken control rods and of a large number of rods with cracks at the twin reactors Oskarshamn 3 and Forsmark 3, no knowledge existed about the details of the failure process (Tinoco and Lindqvist, 2009). It was first when time dependent CFD simulations revealed the structure of the thermal mixing process in the control rod guide tube that the cause, thermal fatigue of the rods, was confirmed and understood (Tinoco and Lindqvist, 2009, Tinocoet al., 2010b, Angele et al., 2011). This subject, together with the general problem of conjugate heat transfer, is analysed in Section 5.

Firstly, this chapter intends to communicate an historical view of CFD applied to the analysis of the mechanical integrity of Light Water Nuclear Reactors. Secondly, the chapter examines and evaluates several methodologies and approaches corresponding to the present and future modelling in the field of thermal hydraulics related to this type of analysis.

2. The determination of loadings

Considering the devastating effect that uncontrolled loadings may have on life and property, this section will begin by evoking the history of pressure vessel advance. This road to progress is paved with disastrous accidents which, during the decade of the eighteen eighties, amounted in the USA to more than 2000 boiler explosions. The establishment of ASME in February 1880 was, according to legend, directly prompted by the need to solve the problem of unsafe boilers (Varrasi, 2005). But it was first after the explosion on March 20, 1905, of the boiler of the R. B. Grover & Company Shoe Factory in Brockton, Massachusetts, (see Fig. 1 below, USGenWeb, 2011, Ellenberger et al., 2004) that the real effort of developing rules and regulations for the construction of secure steam boilers speeded up, urged by the public opinion. During the subsequent years, many states legislated, without too much coordination, about what was considered suitable instructions and procedures, leading to inconsistency in the construction requirements. Finally, in the spring of 1915, the first *ASME Rules for Construction of Stationary Boilers and For Allowable Working Pressures* consisting of one book of 114 pages, known as the 1914 edition, was made available, and the path towards regulation uniformity commenced.

Figure 1. View of the R. B. Grover & Company Shoe Factory before (left) and after (middle and right) the boiler explosion on March 20, 1905, that killed 58 persons and injured 117, (USGenWeb, 2011, Ellenberger et al., 2004).

Since then, the ASME Code has become both larger and more comprehensive, comprising today 28 books, with 12 books dedicated to the Construction and Inspection of Nuclear Power Plant Components. Fortunately, even the boiler explosion trend in the USA has radically changed since 1905, as Figure 2 below shows.

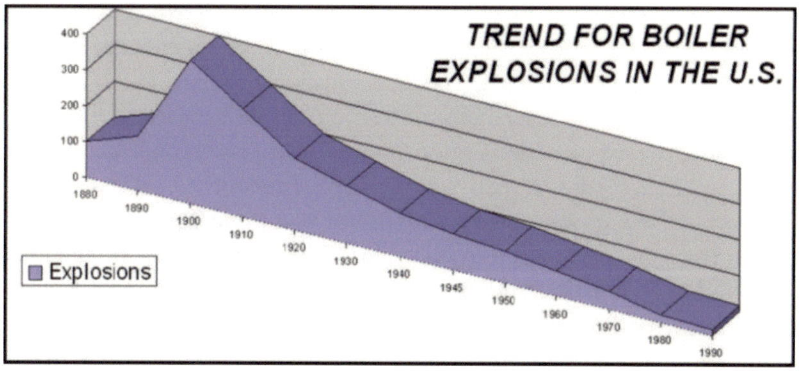

Figure 2. Trend for boiler explosions in the USA (from Hill, 2008).

For nuclear applications, the complete ASME code is mandatory only in the USA. In the European Union, on the other hand, no coordination process has been agreed. The Pressure Equipment Directive, (EU, 1997), has been adopted by the European Parliament and the European Council for harmonizing the national laws of Member States to promote and facilitate trade and exchange between States. But in the nuclear field, the application limits of the Directive and the Nuclear Codes (possibly ASME) have to be agreed with the National Nuclear Regulatory Authorities.

2.1. Section III

The ASME Boiler and Pressure Vessel Code, Section III, Rules for Construction of Nuclear Power Plant Components, of which the last Edition is from 2010, (ASME, 2010), regulates the design and construction of nuclear facility components. This Section "provides requirements for the materials, design, fabrication, examination, testing, inspection, installation, certification, stamping, and overpressure protection of nuclear power plant components, and component and piping supports. Components include metal vessels and systems, pumps, valves, and core support structures."

Section III comprises Divisions 1, 2 and 3 and the introductory Subsection NCA that specifies General Requirements for Divisions 1 and 2. Division 1 contains eight Subsections, namely NB, for Class 1 Components, NC, for class 2 Components, ND, for class 3 Components, NE, for class MC Components, NF, for Supports, NG, for Core Support Structures, NH, for Class 1 Components in Elevated Temperature Service, and Appendices, both mandatory and nonmandatory, including, inter alia, a listing of design and design analysis methods.

2.2. Categorisation of parts in code classes

The aforementioned Code classes are proposed for the categorisation of parts of a nuclear power system, in accordance with the level of importance related to their function in the safe operation of the plant. However, the Code does not provide guidance for classifying the different parts. Classification, which is the responsibility of the Owner, has to be achieved by applying system safety criteria to be found in engineering standards and/or requirements of the Nuclear Regulatory Authorities. For instance, Class 1 components are those that are part of the primary core cooling system, or components that are used in elevated temperature service, and are constructed in accordance with the rules of Subsection NB. Class 2 components are those that are part of various important-to-safety emergency core cooling systems, and are constructed in accordance with the rules of Subsection NC. Class 3 components are those that are part of the various systems needed for plant operation, and are constructed in accordance with the rules of Subsection ND. Class MC components are metal containment vessels constructed in conformity with the rules of Subsection NE. Class CS components are core support structures constructed in accordance with the rules of Subsection NG. Mandatory and/or nonmandatory modifications and/or extensions to these classes may be included by Nuclear Regulatory or other Authorities, (STUK, 2000, IAEA, 2010).

2.3. Mechanical integrity and design specifications

The requirements for mechanical integrity are based on Norms and aim at ensuring that nuclear components shall withstand pressure and other types of loadings without system break or leakage. The Class to which the component belongs to shall govern the choice of Norms that are used for the analysis of its mechanical integrity. Design Specifications for mechanical integrity shall indicate the Loadings and Loading Combinations that a mechanical component is submitted to and, at the same time, the acceptable level of the Loadings and Loading Combinations, i.e. Loading Limits.

Within the realm of the ASME Code, the introductory Subsection NCA covers "general requirements for manufacturers, fabricators, installers, designers, material manufacturers, material suppliers, and owners of nuclear power plants", (ASME, 2010). Here, Subsection NCA-2142 imposes to the Owner or his designee the duty of "identify the Loadings and combinations of Loadings and establish the appropriate Design, Service, and Test Limits for each component or support". For this purpose, Loadings are separated into Design, Service and Test Loadings, and their Limits correspond to the acceptable Loadings permitted in a structural analysis. General directions about the characterization of Design, Service and Test Limits may be found in Subsection NCA-2142.4.

In order to be able to determine the Loadings that a component or support is submitted to, the different operating conditions that may affect the component or support have to be defined. The selection should include Normal Operation, Abnormal Conditions and Accidents that the component or support shall successfully withstand. Also, to be able to estimate the Limits that the Loadings must conform to, each operation condition must be related to an occurrence probability, being a lower probability associated with higher acceptable Limits. This matter may be solved by an event classification such as that of ANSI/ANS (1983a) for

PWR and ANSI/ANS (1983b) for BWR. Here, Plant Condition PC1 corresponds to normal operating conditions, PC2 to anticipated conditions (occurrence frequency of > 10^{-2} times/year), PC3 to abnormal operating conditions (occurrence frequency of 10^{-2} - 10^{-4} times/year), PC4 to postulated accidents (occurrence frequency of 10^{-4}- 10^{-6} times/year) and PC5 to low probability accidents (occurrence frequency of 10^{-6} - 10^{-7} times/year).

2.4. Design loadings and limits

Subsection NCA-2142.1 indicates that the Design Loadings shall be specified by stipulating (a) the Design Pressure, (b) the Design Temperature and (c) the Design Mechanical Loads. External and internal Design Pressure shall be consistent with the maximum pressure difference between the inside and outside of the item, or between any two chambers of a combination unit. The Design Temperature shall be equal to or higher than the expected maximum thickness-mean metal temperature of the part considered. The Design Mechanical Loads "shall be selected so that when combined with the effects of Design Pressure, they produce the highest primary stresses of any coincident combination of loadings for which Level A Service Limits (see below about limits) are designated in the Design Specification". Primary stresses are those arising from the imposed loading, not those developed by constraining the free displacement of the structural system. Primary stresses are necessary to satisfy the mechanical equilibrium of the system. The component or support shall be structurally analysed for the Loadings associated with Design Pressure, Design Temperature and Design Mechanical Loads where, in general, their possible cyclic or transient behaviour is not included.

Design Limits shall designate the limits for Design Loadings. The Limits for Design Loadings shall conform to the requisites given in the applicable Subsection of Section III, i.e. NB, NC, ND, NE, NF or NG. In the event of Loadings that are not structurally analysed, the Limits are set to the same level as Service Limits A (see below).

2.5. Service loadings and limits

From the defined operation conditions, the Service Loadings may now be derived. Service Loadings are all loadings that a component is subject to under predictable normal and abnormal operational conditions and postulated accidents that have to be included in the Design Specification. More specifically, Service Loadings are pressure, temperature loads, mechanical loads and their possible cyclic or transient behaviour (see e.g. Subsection NCA-2142.2). Related loadings shall be integrated in Service Loading Combinations which, together with the corresponding Plant Condition or assigned probability, shall be assessed against relevant Limits.

According to Subsection NCA-2142-3, Service Limits are divided into four levels, namely Service Limits A, B, C and D. Service Limit A corresponds to bounds that contain the safety margins and factors that are required for the component to completely fulfil the specified performance. Service Limit B corresponds to bounds that contain smaller safety margins and factors than Limit A, but that still ensure a component free from damage. Service Limits C and D correspond to bounds with even more reduced safety margins and factors that now may lead to permanent deformation and damage of the component, a situation requiring repair of the component. The last two Limit levels may not be suitable for pressure vessels since their mechanical integrity should not be jeopardised.

The Plant Conditions (operation conditions together with their corresponding occurrence probability) shall decide which Limit to be applied. This coupling shall be account for in the Safety Analysis of the system (see e.g. ANSI/ANS 51.1 or ANSI/ANS 52.1, Appendix 2).

In the evaluation of a component according to Section III, Subsections NB, NC, ND, NE, NF or NG, quantifying values corresponding to Service Limits A to D shall be assigned to the allowed stresses.

2.6. Test loadings and limits

Test Loadings correspond to all Loadings that a component is subjected to during the tests that the component has to undergo. In general, these correspond to pressure tests but, if other types of tests are required, these must be stated in the Design Specification (see Subsection NCA-2142.3).

Test Limits indicate bounds for the tests to be performed on a component. In the evaluation of a component according to Section III, Subsections NB, NC, ND, NE, NF or NG, quantifying values corresponding to the Test Limits shall be assigned to the allowed stresses.

2.7. Load combinations

The overall-categorized Loadings with their corresponding Limits comprise a part of the background necessary to accomplish a structural analysis. Its complementary part is constituted by Load Combinations, the definition of which is, according to NCA-2142, a responsibility of the Owner of the Nuclear Facility, or of his designee. The combination of already identified Loads has to consider the type of Load, i.e. static or dynamic, global or local. In addition, the combination has to determine if they are consecutive or simultaneous, i.e. pressure and temperature loads related to a Loss of Coolant Accident (LOCA). Also, the time history of each Load shall be taken into account in order to prevent an unlikely superposition of non-simultaneous peaks. Eventually, the occurrence probability of each load combination shall be assessed.

Consequently, Design Conditions may include not only static loads like Design Pressure (DP), Design Temperature (TD) and Dead Weight (DW) but also dynamic loads like those generated by, for instance, the opening/closing of one safety valve (GV/SRV(1)). In the combination notation, the number of valves involved is indicated within parenthesis. Also, the occurrence probability of the maximum load may be added as the inverse of the number of valve activations under which this maximum is achieved, for instance "≥ E-3", meaning the maximum load in 1000 activations (see e.g. Table 1).

More specifically, the loadings due to Operational Transients and Postulated Accidents may, for instance, include the Operation Pressure (PO), the Operation Temperature (TO), temperature transient (TT), joint displacement of reactor nozzle (D/RPV), building displacement (DB), condensation-induced water hammer when starting the high pressure Emergency Core Cooling (HP-ECCS) system or the Feed Water System (WH/SC), water hammer by star/stop of a pump (WH/PT) and global vibration due to safety valve discharge (GV/SRV).

The loadings due to Postulated Accidents may also include water hammer caused by a pipe break external to the Containment (WH/PBO), reactor vibrations generated by internal pipe break in the Containment (PRVV/PBI), global vibrations due to condensation oscillations in the suppression pool (GV/CO) and global vibration by condensation-induced pressure pulses, chugging, in the suppression pool, (GV/CH). Finally, loads caused by other incidents may be those corresponding to global vibrations due to Safe-Shutdown Earthquake (GV/SSE).

In a linear structural analysis, the effect of a Load Combinations is evaluated by a simple linear superposition of the response of each normal mode, separately analysed, through the procedure referred to as modal analysis. In cases where the time dependent functions of Loads acting simultaneously are statistically independent of each other, an upper bound of the total response may be estimated by a method denoted Square Root of Sum of Squares (SRSS), and not by the very conservative linear sum of the maxima. If the structural analysis is non-linear, as in Zeng et al. (2011), modal analysis is no longer valid and Collapse-Load analysis, together with non-linear Transient analysis, is required.

Combination	Superposition Rule	Plant Condition	Service Limit Levels
A01	PD + DW	–	Design
A03	PO + DW + TT + D/RPV + D/B	PC1/PC2	A/B
A06	PO + DW + GV/SRV(1)	PC1	Design, A
A07a	PO + DW + GV/SRV(12 ≥ E-3)	PC2	B
A07b	PO + DW + GV/SRV(12 ≥ E-8)	PC4	D
A08a	PO + DW + GV/SRV(12 ≥ E-2)	PC3	C
A09a	PO + DW + WH/SC	PC2	B
A09b	PO + DW + WH/SC	PC3	C
A11	PO + DW + WH/PBO	PC4	D
A13	PO + DW + $[(GV/SVR(1))^2 + (GV/CO)^2]^{1/2}$	PC4	D
A14	PO + DW + $[(GV/SVR(1))^2 + (GV/CH)^2]^{1/2}$	PC4	D
A16	PO + DW + $[(GV/SVR(12 \geq E\text{-}3))^2 + (RPVV/PBI)^2]^{1/2}$	PC4	D
A17	PO + DW + $[(GV/SVR(12 \geq E\text{-}2))^2 + (GV/SSE)^2]^{1/2}$	PC5	D

Table 1. Design Specifications for the feed water system (System 415) of a BWR (Forsmark 1, Forsmarks Kraftgrupp, 2007).

As an example of Design Specifications for pressure- and force-bearing components, Table 1 shows the different Load Combinations corresponding to a part of the feed water system of an internal pump BWR corresponding to Unit 1 of Forsmark Nuclear Power Plant (NPP). In this table, some of the Load combinations appear to be identical but they actually differ in, for instance, the pressure and/or temperature limits. For the sake of simplicity, this information has not been included in the table. It might be of some interest to point out that the Case A17 corresponds to a combination of an earthquake from San Francisco, USA, and another from Sweden, since the spectra of these earthquakes differ in the frequency content. This Combination has a Plant Condition PC5 for the Swedish reality, for which a safe shutdown of the reactor must be guaranteed.

Forsmark is the youngest Swedish NPP that became operational in 1980 with Unit 1 and whose installed power was completed in 1985 with the start of Unit 3. This Unit and Unit 3 of Oskarshamn NPP were the last reactors to be built in Sweden. These two internal pump reactors were designed at the end of the nineteen sixties and beginning of seventies, when the first pocket calculators begun to invade the market. But these calculators were initially quite expensive (Hewlett-Packard's HP-35 cost $395 in 1972 which corresponds today to more than $2000), forcing the structural analyses to the use of rather simplified rough models, such as columns and beams subjected mainly to static loads, computed by means of slide rules. Dynamical effects were considered using a quasi-static approach with adequate safety margins (see e.g. Lahey and Moody, 1993), constituting the only feasible and realistic conservative approach compatible with the existing algorithms and computer capacity of the time. Nevertheless, thanks to the increased knowledge in the physics of single- and multi-phase flows, to the development of numerical models for fluids and structures together with the massive computing power through parallelization and low-price processers, mechanical integrity analysis of Light Water Nuclear Reactors has, during the two last decades, experienced a strong progress.

As discussed above, a nuclear reactor is designed for normal steady-state operation, for normal changes such as start-up, shutdown, change in power level, etc., without going beyond the design limits. In addition, the design must manage expected abnormal conditions and postulated accidents. In this context and according to Lahey and Moody (1993), an accident "is defined as a single event, not reasonably expected during the course of plant operation, that has been hypothesized for analysis purposes or postulated from unlikely but possible situations and that has the potential to cause a release of an unacceptable amount of radioactive material". Consequently, a break in the reactor coolant system has to be considered an accident whereas a fuel cladding tube damage is not.

Among postulated accidents, a group known as Anticipated Transients Without Scram (ATWS), has a preferential status due to the interest that it has attracted since the end of the nineteen seventies. It assumes the partial or total failure of the automatic control rod insertion (SCRAM). A total ATWS has a very low occurrence probability in the reactors of Forsmark NPP. The total of about 170 control rods is divided into independent scram groups, each served by a scram module and containing eight to ten control rods. A scram module consists of a high-pressure nitrogen tank connected by a scram valve to another tank containing water to be pressurized by the nitrogen, water that may act on piston tubes for accomplishing rapid insertion of control rods. It might be of some interest to point out that similar scram systems successfully stopped units 2, 3, 4 and 7 of the Kashiwazaki-Kariwa NPP in Japan at the beginning of the earthquake Chūetsu of July 16, 2007 (the remaining units 1, 5 and 6 and were not operational due to refuelling outage). Still, Forsmark's Units have an independent system of boron injection for emergency stop of the reactors, which is automated for Units 1 and 2 but still manual for Unit 3. Although the automation of this system, which was a Regulatory compliance, is not associated with safety issues, can, nevertheless, lead to new thermal loads for the fuel and/or adjacent structures in the reactor core (Tinoco et al., 2010a). As mentioned above, it is the duty of the owner or designee of the nuclear facility to identify the Mechanical Loadings and/or their Combinations and establish appropriate Limits. It is therefore necessary to analyse the effects of boron injection, a topic to be discussed in somewhat more detail in Section 3 of this chapter.

The design basis accident of an external pump Boiling Water Reactor (BWR) consists of a two-ended instantaneous circumferential break on the suction side of one of the main recirculation lines, resulting in a discharge from both ends. In a BWR with internal pumps, this postulated accident is not possible since there are no main recirculation lines. For an internal pump BWR, there no longer exists a single design basis accident, but rather a series of accidents, whose consequences for safety and mechanical integrity are at least an order of magnitude smaller than the rupture of a main recirculation line. One of these accidents consists of the two-ended instantaneous circumferential break of a steam line, which produces the highest loads in the vessel and reactor internals. A similar break of a feedwater line will result in lower Loads due to damping two-phase effects, in this case mainly from rapid boiling, i.e. flashing. Therefore, a design basis accident to be analysed for an internal pump BWR is a steam line break that will be discussed in more detail in the following Section.

3. A steam line break

Until the mid-nineties, studies and analysis of the effects of a steam line break were mainly based on experiments such as those of The Marviken Full Scale Critical Flow Tests (1979) where the pressure vessel consisted of a cylindrical tank with no internal parts, filled only with steam. For this reason, the only pressure waves considered in the loading analyses were the alternating decompression and compression waves formed in the pipe segment coupled to the cylindrical tank, between the break section and the coupling section to the tank (see for example Laheyand Moody, 1993, pp. 498-501). Consequently, the Loads given by the standard ANSI/ANS-58.2-1988, (ANSI/ANS, 1988), contain a component due to pressure waves of slightly over 30% of the product of the initial pressure of the cylindrical tank times the area of the section break. That the presence of the concentric steam annulus between the reactor vessel and the core shroud has an important part in the process of decompression by supporting standing pressure waves, (Tinoco, 2001), has been understood after the CDF analysis to be briefly describe in what follows.

A three-dimensional numerical analysis of the rapid transient corresponding to the rupture of the steam line (Tinoco, 2002) started in the late nineties with the purpose of obtaining the time signal of the pressure inside the reactor pressure vessel and its internals, to be later used in a structural time history analysis. This work was motivated as a part of the structural analysis related to replacement of the core grid of Units 1 and 2 of Forsmark NPP from the original, fabricated by welded parts that had developed cracks in the welds, to a new manufactured in one piece.

Historically, the three-dimensional numerical simulation of the fast transient was one of the first successes in this area of thermal-hydraulics. This was due to several factors that allowed an analysis with relatively limited computational resources, namely a negligible effect of turbulence, single-phase treatment due to flashing delay and the lack of importance of friction allowing relatively coarse mesh (see Tinoco, 2002).

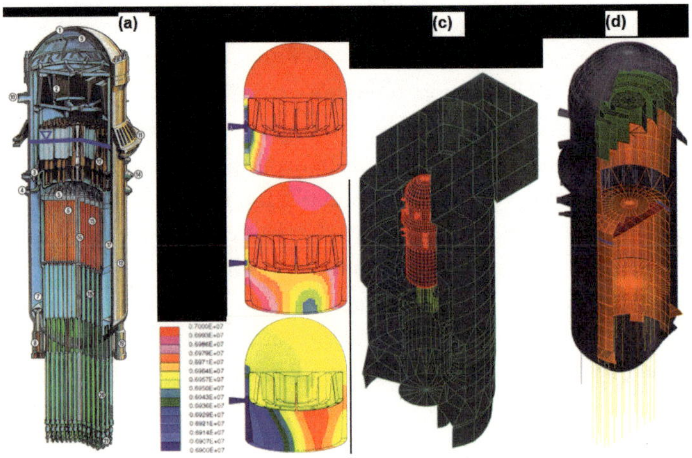

Figure 3. (a) View of the reactor with water surface indicated by blue line, (b) Model of the steam dome, annulus and dryers (down to water surface) after steam line break: 9 ms (upper view), 20 ms (central view) and 100 ms (lower view), (c) Simplified FEM model of the reactor with containment building, (d) detailed FEM model of the reactor with core grid in red.

Figure 3 shows in (a) an image of the reactor with the water surface indicated by a blue line. In (b) this figure shows a sequence of views of the pressure on the inner surface of the numerical model CFD ("Computational Fluid Dynamics") model (Tinoco, 2002), containing the steam dome, the annular region of steam and steam dryers. The steam separators, located under the dryers were modelled together as a single barrel without details. The first view of the sequence corresponds to 9.2 ms after the break, where the decompression wave extends with a single front symmetrically with respect to the axis of the steam nozzle, on whose outer end the line break is assumed to be located. The second view corresponds to 20.2 ms after the break, showing a picture of a less ordered and more chaotic pressure field due to reflections with different parts of the complex geometry of the reactor vessel. The third view corresponds to 100 ms from the break, displaying a more simplified image of the pressure than in the previous view, due mainly to the general decompression of the reactor that has almost homogenised the dome pressure. Image (c) shows a general view of the low-resolution structural analysis FEM model, including the containment building. Image (d) shows a view of the high-resolution FEM model of the reactor.

A structural analysis was performed using both models because, due to the rupture of the steam line, there is an effect, external to the reactor vessel, on the reactor containment building, not considered in Tinoco (2002) since it only treats internal effects. The analysis was conducted using the method of direct time integration, (Hermansson and Thorsson, 1997). The time signals of the loads were applied to the structures and, as a result, response spectra were generated for the different structures. A response spectrum is a graph of the maximal response (displacement, velocity or acceleration) of a linear oscillatory system (elastic structure) forced to move by a vibrational excitation or pulse (shock). Figure 4 shows in (a) two time signals of load components in the direction of the axis of the steam nozzle with the line break, acting on a node located in the core grid. The lower value corresponds to that computed directly by the CFD model. The higher, amplified value takes into account the effect of the second end of the break, which is connected to the adjacent steam line (Hermansson and Thorsson, 1997). Due to this connection in pairs of the steam lines, the reactor vessel is also decompressed, with a certain delay, through the contiguous steam nozzle. In (b),Figure 4 shows the smoothed envelope spectra in the direction of the axis of the steam nozzle with the line break, corresponding to the same node. It might be of same interest to reiterate that the loads applied in this analysis are about twice the loads according to standard ANSI/ANS-58.2-1988, (ANSI/ANS,1988), due to an effect of resonance of pressure waves in the steam annulus (Tinoco, 2001).

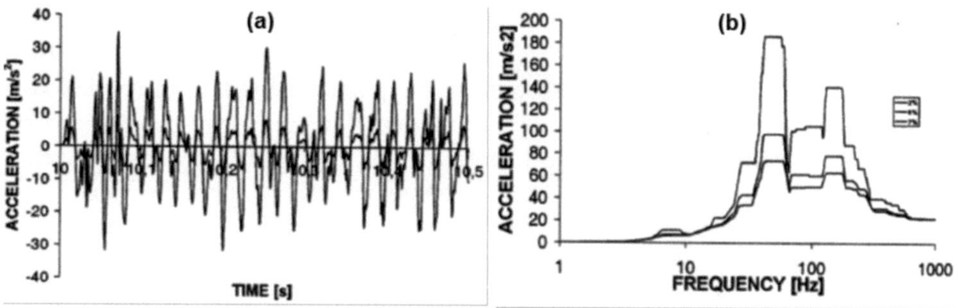

Figure 4. (a) Time signals of the acceleration of a node in the core grid in the direction of the axis of the steam nozzle with break: value calculated by CFD and amplified value. (b) Smoothed envelope spectra in in the direction of the axis of the steam nozzle with break for the same node in the core grid.

4. The problem of mixing

Mixing the same fluid with different properties like temperature or concentration, or with different phases like steam and water, or even different fluids, is a far from trivial industrial problem. The process involved in the operation of Light Water Reactors contains many situations where mixing is of crucial importance like the mixing of reactor water and feedwater in BRWs, the mixing needed to regulate boron concentration in PWRs, etc. Yet, a high degree of mixing between the originally rather segregated components may be a hard task or, depending on the space and time scales of the mixing process, a complete mixing may even be impossible without special mixing devices. For instance, specially designed spacers in the nuclear fuel are needed to enhance the turbulent heat transport, and mixers are used to avoid thermal fatigue due to temperature fluctuations when flows with different temperatures meet within a tee-junction.

All these situations relay on turbulent mixing that, according to Dimotakis (2005), "can be viewed as a three-stage process of entrainment, dispersion (or stirring), and diffusion, spanning the full spectrum of space-time scales of the flow". The mixing level, or quality of mixing, may be estimated, according to Brodkey (1967), through two parameters. The first parameter is the scale of segregation that gives a measure of the size of the unmixed lumps of the original components, i.e. a measure of the spreading of one component within the other. A length scale or a volume scale, analogous to the integral scales of turbulence, may be defined based on the Eulerian autocorrelation function of the concentration. Since the scale of segregation provides a measure of the extent of the region over which concentrations are appreciably correlated, it constitutes a good assess of the large-scale (turbulence-controlled eddy break-up) process, but not of the small scale (diffusion-controlled) process. The second is the intensity of segregation, given by the one-point concentration variance normalised with its initial value. It constitutes a measure of the concentration difference between neighbouring lumps of fluid, i.e. it describes the effect of molecular diffusion on the mixing process.

But, according to Kukukova et al. (2009), a general, objective and complete definition of mixing, or of its opposite, segregation, needs a third parameter, or dimension, as the authors of the reference express it, namely the exposure or the potential to reduce segregation. This rate of change of segregation associated to a time scale of mixing, combines the strength of the interaction, in this case the turbulent diffusivity, the concentration gradients and the contact area between components. Therefore, good mixing characterized by a small scale of segregation and a low intensity of segregation may be achieved if the mixing process is contained within the exposure time scale, i.e. for adequately high exposure.

However, many design problems of light water reactors that have arisen during their operation, originate from overconfidence about the mixing properties of turbulence. A classic example is that of the mixing of feedwater and reactor water in the downcomer of a BWR, in connection with Hydrogen Water Chemistry (HWC) to mitigate Intergranular Stress Corrosion Cracking (IGSCC) of sensitized stainless steel. The injection of reducing hydrogen gas to the feedwater was intended for recombination with oxidizing radicals contained in the reactor water and produced by radiolysis, through a turbulence-induced, optimal mixing in the downcomer. Surprisingly, the recombination effect for Units 1 and 2 of Forsmark NPP was not in proportion to the amount of injected hydrogen and the only solution appeared to be a major increase in the addition rate of hydrogen. But, in 1992, a very primitive and rough CFD model of an octant (45º) of the downcomer (Hemström et al., 1992) indicated that the flow field and the mixing were rather complex and highly tree-dimensional, leaving the regions between feedwater spargers with practically no mixing, and that buoyancy ostensibly aggravated the mixing issue. The poor mixing in the downcomer of Units 1 and 2 of Forsmark NPP arose from the design of the core shroud cover that, radially and freely over its entire perimeter, directed the warmer reactor water coming from the steam separators into the downcomer. Owing to the limited azimuthal extent of the four feedwaterspargers, four vertical regions without direct injection of colder feedwater were established in the downcomer. Due to buoyancy combined with mass conservation, the colder and falling regions contracted azimuthally, making the warmer and slower regions to expand. Turbulence acted within the colder regions, tending rather effectively to achieve temperature homogeneity, and at the boundary areas between the warmer and colder regions. The transit time of the vertical flow in the downcomer was not long enough for turbulence to achieve a complete temperature homogenization of the recirculation water. In other words, the turbulent diffusivity, the temperature gradients and the contact surfaces between cold and warm water were not large enough to achieve a good mixing, i.e. the exposure was too low. The mixing was achieved further downstream in two main steps, namely by forcing cold and warm water to meet within the inlet pumps, at the bottom of the downcomer, and by rotating agitation within the remnant vortexes at the pump outlets, produced by the Main Recirculation Pumps (MRP). This mixing process is quite effective, leaving only less than 3 % difference in temperature (Tinoco and Ahlinder, 2009) but for HWC, this homogeneity comes too late in the process since radiation at the level of the core in the downcomer is needed for triggering recombination of oxygen with hydrogen.

Finally, the problem of poor mixing in the downcomer was significantly reduced by modifying the design of the core shroud cover. Cylindrical-vertical walls were placed at the edge of the cover in the azimuthal locations facing the regions between spargers in order to completely block the reactor water flow. Radial flow out of the cover was allowed only regions in front of the spargers. This design was first tested in a rather advanced CFD model (Ahlinder et al., 2007), used to investigate HWC. The results obtained showed a higher degree of recombination with the modified core shroud. However, the decisive factor in the recombination was the level of radiation absorbed by the water in the downcomer, indicating that an effective recombination might not be possible over the complete fuel cycle without a modified core design. For a couple of years ago, this shroud cover design was implemented in Units 1 and 2 of Forsmark NPP, as a step in the modernisation and improvement of the process and safety to cope with a programmed power uprate. HWC was not carried out due to the aforementioned core design issue but an improved mixing in the downcomer has other advantages like minimizing the risk of cavitation in the MRPs.

The most complete CFD model including for the first time rotating MRPs was developed in 2008 (Tinoco and Ahlinder, 2009) for studying thermal mixing in an internal pump BWR. Figure 5 shows in (a) once more the view of the reactor as a reference, and in (b) it shows a view of the geometry involved in the CFD model that is limited from above by the free water surface. The model includes the downcomer (the water annular region between the inner surface of the reactor and the outer surface of the core shroud), the MRPs (4 pumps for this half of the reactor) and the lower plenum with the guide tubes of the control rods which extend to the entrance of the core reactor. In (c) a MRP is shown in more detail and in (d) the lower plenum, also in more detail. The complete CFD model contains slightly more than 25 million cells to represent the included part of the reactor. The correct flow through the model is achieved by rotating the pumps to the adequate rotational speed for normal operation. The thermodynamic properties of water as a function of pressure and temperature have been included in the physical properties of the studied fluid, allowing with that the analysis of the mixing of warmer recirculation water with cooler feed water. The results indicate a substantial mixing at the lower plenum inlet, being the maximum difference in the temperature of water entering the reactor core of about 2.5 ºC (see Tinoco and Ahlinder, 2009). This model may be used in the future to analyse the mixing of boron injected to the reactor with the surrounding water, as suggested below.

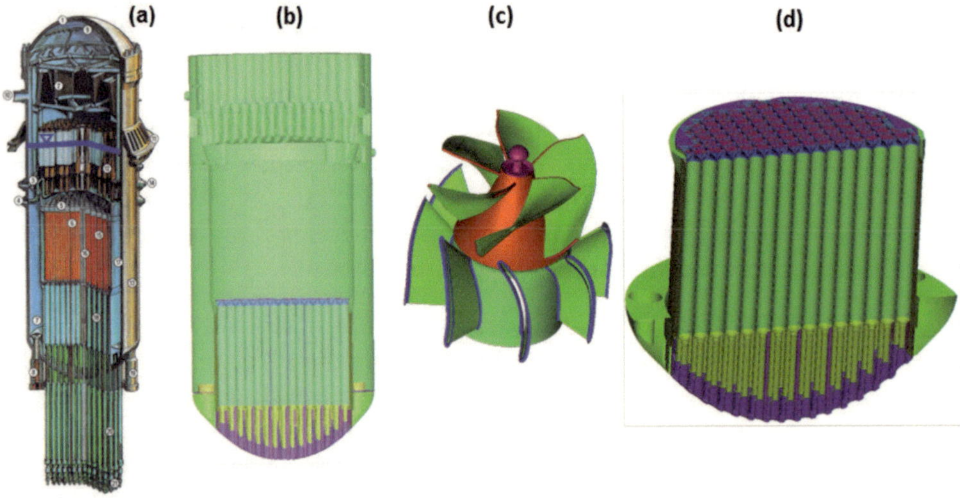

Figure 5. (a) View of the reactor; (b) Model of the Downcomer, MR Pumps and Lower Plenum limited from above by the free water surface; (c) Model of MR Pump with diffuser vanes; (d) Lower Plenum detail with control rod guide tubes.

Originally, this CFD model consisted of two separated sub-models linked by boundary conditions at a common surface constituted by the Lower Plenum Inlet, since this split was geometrically and computationally optimal. The Lower Plenum Inlet is constituted by cylindrical-rectangular openings located almost directly downstream of the MRP outlets equipped with diffuser vanes. Even if these diffuser vanes recover an important amount of the mechanical energy stored in the rotation of the flow imparted by the pump, a vortex is still formed with a significant downstream extension, being this vortex responsible for an appreciable part of the thermal mixing. The use of a surface, as outlet for the first sub-model and as inlet for the second sub-model, to transfer the information of the flow behaviour from one sub-model to the other is not adequate. The information corresponding to the dif-

http://dx.doi.org/10.5772/52691

ferent fields is correctly transferred but no information about the field tendencies, i.e. the different gradients, reaches the second sub-model. In this case, the further development of the vortices coming from the pumps is corrupted beyond the inlets of the second sub-model. The result is a rather strong underestimation of the thermal mixing, predicting temperature differences of the order of 10 %, (see Tinoco and Ahlinder, 2009), temperature differences that have never been detected by measurements or, indirectly, by contingent fuel damage. The analysis of the different turbulence parameters objectively endorses this difference in mixing, which leads to the only reasonable recommendation of using a larger and computationally heavier model but right from the point of view of the physics of the flow. Model separations in CFD simulations, similar to the one described above, will deliver proper results only if the different gradients associated with the flow are small or negligible, otherwise a finite, common, transition volume is needed. This type of coupling between two CFD models might be difficult to achieve in rather standard commercial codes.

By Regulatory compliance, the safety system for the injection of boron to achieve an emergency stop of the reactor in case of an ATWS is now automatically operated in Units 1 and 2. Due to the automation, the accidental or programmed injection of boron will be associated with a set of new cases of postulated accidents that have to be analysed. Since the injection of boron in these Units is done in only two points, closely and not symmetrically located in the upper downcomer, the probability of an inhomogeneous distribution of boron, especially at the entrance of the core, is very high, as a preliminary study using a reduced model containing only the downcomer indicates (Tinoco et al., 2010a). This situation may lead to rapid variations in space and time of the reactivity of the core that might result in local maxima of thermal power that could ultimately damage the fuel and/or adjacent structures. The CFD simulations of boron injection and its mixing with the reactor water have to be linked to three-dimensional simulations of the neutron kinetics of the reactor core in order to correctly capture the physics of these thermal power variations.

5. Heat transfer

Turbulent thermal mixing implies, of course, transfer of heat from the warmer component, or fluid, to the colder component, resulting in a tendency to temperature and enthalpy homogenization. But, depending on the geometry of the thermal system, the process towards homogenization may have a rather well defined composition that involves the system boundaries which, in general, correspond to solid structures. These structures interact thermally with the fluid or fluids involved in the thermal mixing process, being the composition of the process decisive for the behaviour of the structures. According to the viewpoint of this work, the composition of the thermal mixing process includes firstly the morphology of the turbulent flow involved, i.e. the flow geometry, the eddy structure, the presence of stable boundary layers, etc. Secondly, it includes the thermophysical properties of the fluid and the thermophysical and geometrical properties of the solid boundaries. Depending on the composition of the thermal mixing process, the solid boundaries may or may not be exposed to the problem of high-cycle thermal fatigue. Therefore, the composition of the thermal mixing process, as defined above, implies the analysis of the complete system, including the solid boundaries, i.e. the type of heat transfer which is denoted as conjugate heat transfer, with rather advanced requirements regarding comprehensive information of the turbulent flow involved. The aforementioned concepts discussed so far will be emphasized through a slightly different analysis of the paradigm of thermal mixing and fatigue, i.e. the tee-junction.

5.1. Thermal mixing in a tee-junction

This exemplar of thermal mixing has been and is still being extensively studied, both experimentally and numerically, as the work of Naik-Nimbalkar et al. (2010) confirms with a rather wide, but not absolutely comprehensive, review of the literature about the subject. Probably the review should be completed by mentioning the experimental work of Hosseini et al.(2008) about the classification of the branch jet flow and about the structure of the flow field (Hosseini et al., 2009). Also other publications not included in the aforementioned review, (Muramatsu, 2003, Ogawa et al., 2005, Westin et al., 2008, OECD/NEA, 2011), that are experimental or including an experimental part for validation of a numerical simulation, investigate the bulk flow structures of the mixing in a tee-junction. All the references mentioned in this Subsection, and many others not included due to lack of space, have, as the main motivation, the purpose of studying, understanding and eventually controlling the wall temperature fluctuations associated with thermal fatigue. Particularly the reference OECD/NEA (2011), which is a rather major effort to validate numerical simulations of the thermal mixing in a tee-junction, clearly proclaims: "In T-junctions, where hot and cold streams join and mix (though often not completely), significant temperature fluctuations can be created near the walls. The *wall temperature fluctuations* can cause cyclical thermal stresses which may induce *fatigue cracking*". But none of these references presents temperature or velocity data of the wall near region, i.e. the boundary layers, or of temperature at the walls. Even worse, the solid walls are never involved in any measurements, although it is the propagation of temperature fluctuations from the surface and inside the solid that may lead to fatigue cracking.

The reason for the absence of this crucial part of the mixing matter may be the great difficulty in measuring very close to walls and within the walls. For instance, velocity measurements in a thin boundary layer with Lase Doppler Velocimetry (LDV) are biased by even small density differences in the fluid, and temperature measurements within the solid are intrusive, cannot be relocated and may lead to errors due to contact continuity. But, in situations when temperature may be considered as a passive scalar, i.e. relatively small temperature differences not affecting the flow, the velocity measurements may be carried out in isothermal conditions. Since measurements of this kind have been done in this and other situations, to be reported in what follows, the true reason for the absence of this type of measurements in the study of tee-junction flows may be lack of knowledge about how vital the information related to temperature fluctuations close to the wall and within the wall is for understanding and being able to estimate the risk for wall thermal fatigue in a specific case. Of course, the information about flow behaviour close to the wall is directly related to the behaviour of the flow in the bulk, but the physical properties of fluid and the physical and geometrical properties of the wall will also affect the near wall flow, as the next paragraphs of this Subsection will describe. In other words, the problem of turbulent thermal mixing in a tee-junction is a process with a well-defined composition that has to be solved in its wholeness, i.e. a conjugate heat transfer problem.

As the preceding discussion reveals, measurements of the temperature fluctuations due to heat transfer within the boundary layer of a fluid close a wall and/or within the solid wall are very scarce. The work of de Tilly et al., (2006), de Tilly and Sousa (2008a) and de Tilly and Sousa (2008b) deals with the development of a multi-thermocouple planar probe of dimensions of the order of 300 μm to measure the instantaneous values of the wall heat flux in a boundary layer. Preliminary results of the wall heat flux have been obtained for the boundary layer flow developing downstream of a two-dimensional tee-junction. Even if this device is very promising, some questions arise from these tests, namely how to check the accuracy of the measurements with an independent method, how to handle thinner boundary layers than the instrument and how to deal with three-dimensional flow. In any event, this novel concept deserves a positive and successful development as a significant contribution to a difficult but important subject.

The measurements reported by Mosyak et al. (2001) seem to move this discussion away from the trail of wall temperature fluctuations and the risk of fatigue cracking since this reference treats the subject of constant heat flux boundary condition in heat transfer from the solid walls of channel flow. However, the experimental data substantiate the analytically demonstrated fact (Polyakov, 1974) that temperature fluctuations near a wall differ for different combinations of fluid and solid in conjugate heat transfer problems. Also Direct Numerical Simulations (DNS) of conjugate heat transfer (Tiselj et al., 2001, Tiselj et al., 2004, Tiselj, 2011) corroborate this information, where the governing parameters of the fluid-solid combination are the wall thickness d_w, the thermal activity ratio $K=\sqrt{(\rho \cdot c_p \cdot \lambda)_f / (\rho \cdot c_p \cdot \lambda)_w}$ and the ratio of thermal diffusivities $G=\alpha_f / \alpha_w$. Here, ρ is the density, c_p the isobaric specific heat, λ the thermal conductivity and $\alpha=\lambda / (\rho \cdot c_p)$ the thermal diffusivity. The sub-indices f and w indicate respectively "fluid" and "wall". Depending on the values of d_w and K, two asymptotic limits of the conjugate turbulent heat transfer exist at the same Reynolds and Prandtl numbers and at the same wall heat flux. When $K \rightarrow 0$ and $d_w > 0$, the turbulent temperature fluctuations do not penetrate the wall and the wall temperature does not fluctuate. When $K \rightarrow \infty$, d_w is of no importance, the turbulent temperature fluctuations do penetrate the wall and the wall temperature does fluctuate. In a particular solid-fluid case with given K and d_w, the wall temperature behaviour will be between these two limiting cases, as may be observed through the results of Tiselj et al. (2001), given in Figure 6.

The results displayed in Figure 6 are not completely general since they have been obtained for $G=1$, as Tiselj and Cizelj (2012) point out, while $Pr \approx 0.9$, $K \approx 0.2$ and $G \approx 0.04$ for water and stainless steel at the pressure and temperature conditions of a BWR. Nonetheless, the main conclusion of these results is that, despite the large spectrum of temperature fluctuation variations as functions of the aforementioned material and geometrical parameters, the heat transfer rate for the limiting case $K = \infty$ is only 1 % higher than that for the other limiting case $K = 0$. The reason for this coincidence is the fact that the heat transfer rate is associated with the mean temperature gradient, which is very similar in both cases. These are good news for the heat exchanger designers, but the large spectrum of temperature fluctuation variations is bad news for the nuclear and other designers responsible for the mechanical integrity of solid structures.

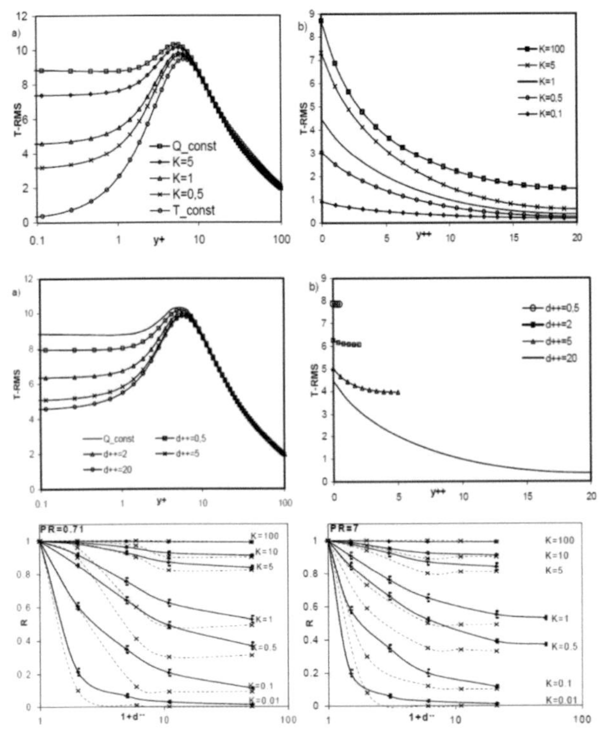

Figure 6. Temperature fluctuations from Tiselj et al. (2001) with ASME permission. Upper view: RMS-values, Pr = 7, d^{++} = 20, (a) fluid; (b) solid. Middle view: RMS-values, Pr = 7, K = 1, (a) fluid; (b) solid. Lower view: RMS-values normalized with ideal RMS-value of $K = \infty$, (a) Pr = 0.71; (b) Pr = 7; dotted lines from Kasagi, N., Kuroda, A., and Hirata, M., (1989).

The value $G=1$ associated with the results of Tiselj et al. (2001) shown in Figure 6, implies that the scale $y^{++} \equiv y^{+}$, $d^{++} \equiv d^{+}$ since $y^{++} = \sqrt{G}y^{+}$, a scale used for the results of Figure 6. Tiselj and Cizelj (2012) mention that this scale has an uncertain meaning and should be abandoned. Also, smaller values than $G=1$ will probably lead to lower temperature fluctuations inside the wall, as the results of Tiselj and Cizelj (2012) indicate, but for $Pr=0.01$. According to Figure 6, an increase in the Prandtl number will further decrease, though moderately, the temperature fluctuations inside the wall. The conclusion of Tiselj and Cizelj (2012) is that, for the combination liquid sodium-stainless steel ($K \approx 1$, $G \approx 10$), the results indicate a relative high penetration of turbulent temperature fluctuations into the simulated heated wall.

http://dx.doi.org/10.5772/52691

The lower values of the parameters for the combination water-stainless steel of BWR suggest that the penetration of turbulent temperature fluctuations in walls will be rather low under the conditions investigated through DNS simulations of channel flow. But, in other circumstances such as tee-junctions or control rod stems, which will be discussed in the next Subsection, the temperature fluctuations seem to be large enough to cause damage through thermal fatigue. It seems apparent that further investigation is needed to elucidate the wall behaviour in cases where no stable boundary layers are built and large structures coming from the bulk flow may strongly interact thermally with the walls. A first step in this direction has been taken by van Haren (2011), where a DNS of a tee-junction has been carried out. Unfortunately, only very preliminary results have been reported since, due to high computer requirements, the run-time was not enough to obtain a statistically stationary and converged solution.

Hopefully, the preceding discussion has made clear that the problem of thermal fatigue induced by flow temperature fluctuations is a conjugate heat transfer problem, i.e. a heat transfer problem of the specific combination fluid-wall. More knowledge is needed to understand the aforementioned interaction between large structures generated in the bulk flow and the walls. This knowledge should be obtained primarily by experiments but further difficulties with temperature measurements should be added to those already mentioned, i.e. intrusive, non-relocatable and contact continuity problems. For instance, the specificity of the combination fluid-wall questions the use in an experimental facility of Plexiglas or other material than the precise of the prototype to be investigated. Even the conditions of the experiment should match those of the prototype since they affect the different material parameters that ultimate influence the thermal fluctuations. Intrusive measurements have to be done with material properties matching those already mentioned of the walls, not only thermal conductivity. For instance, at the FATHERINO 2 facility at CEA Cadarache, briefly described by Kuhn et al. (2010), a thin brass mock-up ("Skin of Fluid Mock-up") has been developed to visualize, by infrared thermography, the mean and fluctuating temperature fields at the interface of the fluid and the wall. This mock-up may only have relevance as a case for validating numerical simulations, as it is used by Kuhn et al. (2010), but more knowledge is needed about the correct conditions, to be discussed below, for simulating the wall heat transfer, including the right distribution of temperature fluctuations within the wall.

Numerical simulations of conjugate heat transfer including temperature fluctuations inside the solid structure other than DNS have already been carried out using diverse approaches. The previously mentioned reference, Kuhn et al. (2010), uses a LES approach to resolve the flow and temperature fields inside the fluid for two cases, namely a thin-wall brass model and a thick-wall steel model. The results show relatively large differences between models in both the mean temperature fields and the distribution of temperature fluctuations inside the solids. This is attributed only to the heat conduction in the thicker wall, apparently both radial and axial, with no further analysis of other possible reasons, i.e. nothing is commented about the different combinations fluid-solid. From the numerical point of view several questions arise, firstly the fact that the same grid is used for the two walls. Secondly, y^+ may be up to 5 in the mixing region when requirements in LES of channel flow, for less than 1 % difference in heat transfer rate with DNS results, are $y^+ = 0.5$, $x^+ = 30$ (streamwise) and $z^+ = 10$ (spanwise), according to Carlsson and Veber (2010). In Kuhn et al. (2010) nothing is mentioned about the grid resolution in the streamwise and/or spanwise directions. However, a validation like that of Carlsson and Veber (2010) about the distribution of temperature fluctuations has not yet been carried out, a situation that sheds some doubts about the accuracy of the results obtained in Kuhn et al. (2010) and, for the rest, of the results obtained with any other methods different from DNS.

A brief and definitely incomplete review of other approaches reveals the use of LES with thermal wall functions (Châtelain et al., 2003). The results show that the temperature fluctuations in the vicinity of the wall and inside the wall can be highly underestimated a fact that is corroborated by Pasutto et al. (2006) and Jayaraju et al. (2010). Also some attempts have been done with the RANS approach, with varied results (Keshmiri, 2006, Tinoco and Lindqvist, 2009, Tinocoet al., 2010b, Craft et al., 2010), and with the more unconventional Lagrangian approach of Pozorski and Minier (2005). The majority of the numerical results discussed in this Subsection are waiting for an experimental validation, or numerical by DNS, in order to be able to establish the correct methodology to be used in future simulations for estimating the risk of thermal fatigue.

5.2. The control rod issue

Broken control rods and rods with cracks were found in the twin reactors Oskarshamn 3 and Forsmark 3 in the fall of 2008.As a part of an extensive damage investigation, time dependent CFD simulations of the flow and the heat transfer in the annular region formed by the guide tube and control rod stem were carried out with a model containing 9 million computational cells (Tinoco and Lindqvist, 2009). Due to limitations in time and computer resources, only a quadrant of the guide tube and control rod with an axial length of approximately 1 m was included in the model, as Figure 7 shows. Also an Unsteady RANS (URANS) approach with the SST k-ω model was chosen for handling turbulence.

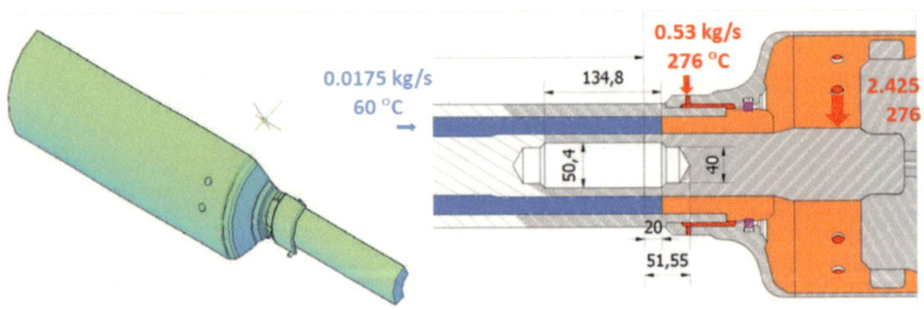

Figure 7. CAD view (left) and drawing of a part of the CFD model used inTinoco and Lindqvist (2009).

In this figure, where gravity is from right to left in the right view, the different mass flow rates per quadrant are indicated with their temperatures by arrows. The upper bypass inlet (right) divides its mass flow rate into two holes of14.6 mm in diameter, resulting in inlet velocities of about 10 m/s. The lower bypass inlet has only one hole of 8 mm in diameter that connects to an annular channel before the flow is delivered with relatively low velocity to the annular region between control rod stem and guide tube.

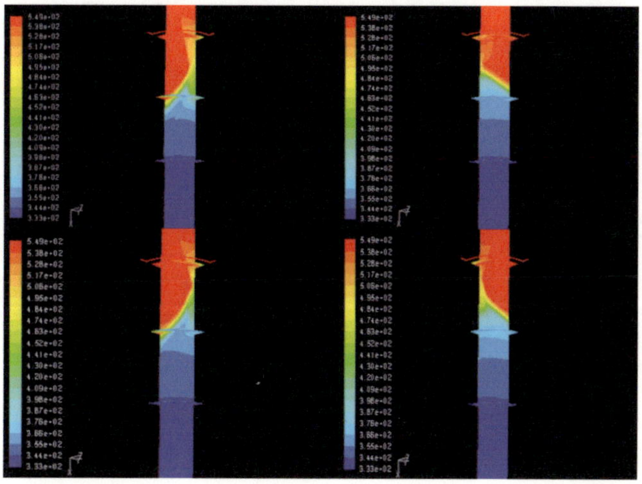

Figure 8. Temperature sequence showing a large scale variation of the surface temperature of the control rod stem with a period of ~15 seconds (temperatures in K).

Figure 8 shows a typical time sequence, from left to right and in two rows, of the surface temperature of the control rod stem with a large scale variation with a frequency of approximately 0.07 Hz. Three horizontal planes are shown in the views, namely the upper plane corresponding to the lower bypass inlet, the middle plane corresponding approximately to the region of the break (10 cm below the lower bypass inlet) and the lower plane corresponding approximately to the end of the inlet region of the cold crud-cleaning flow (20 cm below the lower bypass inlet). These simulations together with metallurgical and structural analyses indicated that the cracks were initiated by thermal fatigue.

The results of the damage investigation were accepted by the Swedish Nuclear Regulatory Authorities for a reactor restart with new control rods. Meanwhile, further studies had to be accomplished to clarify some remaining matters. One of the contested issues concerned the validity and accuracy of the CFD simulations. Consequently, new CFD models, (Tinoco et al., 2010b), were developed in conformity with the guidelines of Casey and Wintergerste (2000) and Menter (2002), while the URANS approach with the SST k-ω model was preserved, and validation experiments were carried out. The largest CFD model contained 16.8 million cells because large effort was put into maintaining a value of y^+~ 1 for the dynamic mesh resolution in the near wall region of the control rod stem at the level of the mixing area. Validation tests had indicated that a y^+ close to 1 is a necessary condition to ensure sufficiently accurate heat transfer results (Tinoco et al., 2010b), but no control was done about the values of x^+ and z^+ since the results of Carlsson and Veber (2010) were at that time not known. Figure 9 shows the low y^+ values for the grid at the stem wall in the mixing region indicated by the horizontal black line for different times of the simulation.

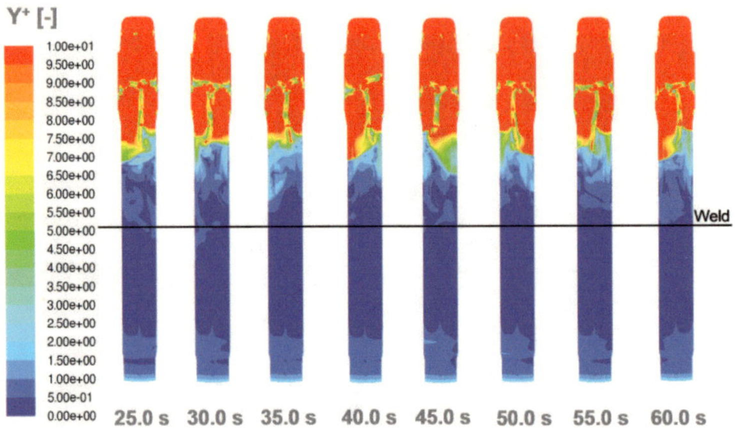

Figure 9. Grid y^+ adjacent to the control rod wall at different times of the simulation.

Figure 10 shows in the upper view a sequence of colour pictures representing the velocity magnitude, maximized to 1 m/s (red scale ≥ 1 m/s), in a vertical plane bisecting the quadrant and containing the region below the jets of the upper bypass inlet. The pictures are completely symmetric since the original data of the quadrant have been mirrored with respect to vertical axis. In this sequence, it is possible to observe many structures forming and moving downwards, towards the lower bypass inlet, in the annular region between the stem and the guide tube. In this view, the black line indicating the level of the stem weld corresponds to the level of plane 7 shown in the middle view.

The middle view depicts five time signals of the axial vertical velocity component of the fluid at radially 2 mm from the stem wall. The points are located in plane 7 (blue line in left reference picture) at 5 different azimuthal positions. The signals indicate large vertical intermittent movements downwards, with speeds of the order of 1 m/s and frequencies of 0.1-0.2 Hz which corroborate the impression given by the pictures of the upper view. This is the dynamical mechanism of the mixing responsible for the thermal fatigue of the control rod stems. This effect has been further confirmed by simulations of guide tubes without the upper bypass inlet, reported in Tinoco and Lindqvist (2009), where no such structures are present.

The lower view shows the thermal consequences of the above described dynamical process that transports large lumps of warm fluid downwards with relatively large velocities. The maxima of the temperature time signal correlate well with the minima of the axial vertical velocity component. The fact that the temperature signal (in blue) corresponds to only one single fluid point indicates that the lumps of warm fluid cover a large part of the annular gap modelled in the quadrant. This view also shows the temperature signal of two points located inside the solid, namely the first located radially at 0.5 mm from the stem outer wall

http://dx.doi.org/10.5772/52691

(in green) and the second located radially at 2 mm from the wall (in red). In this graph, it is possible to observe that, compared to the fluid signal, both solid signals are damped, with high frequency variations filtered out. Also, the red signal is not always smaller than the green indicating switches in the direction of the heat flux.

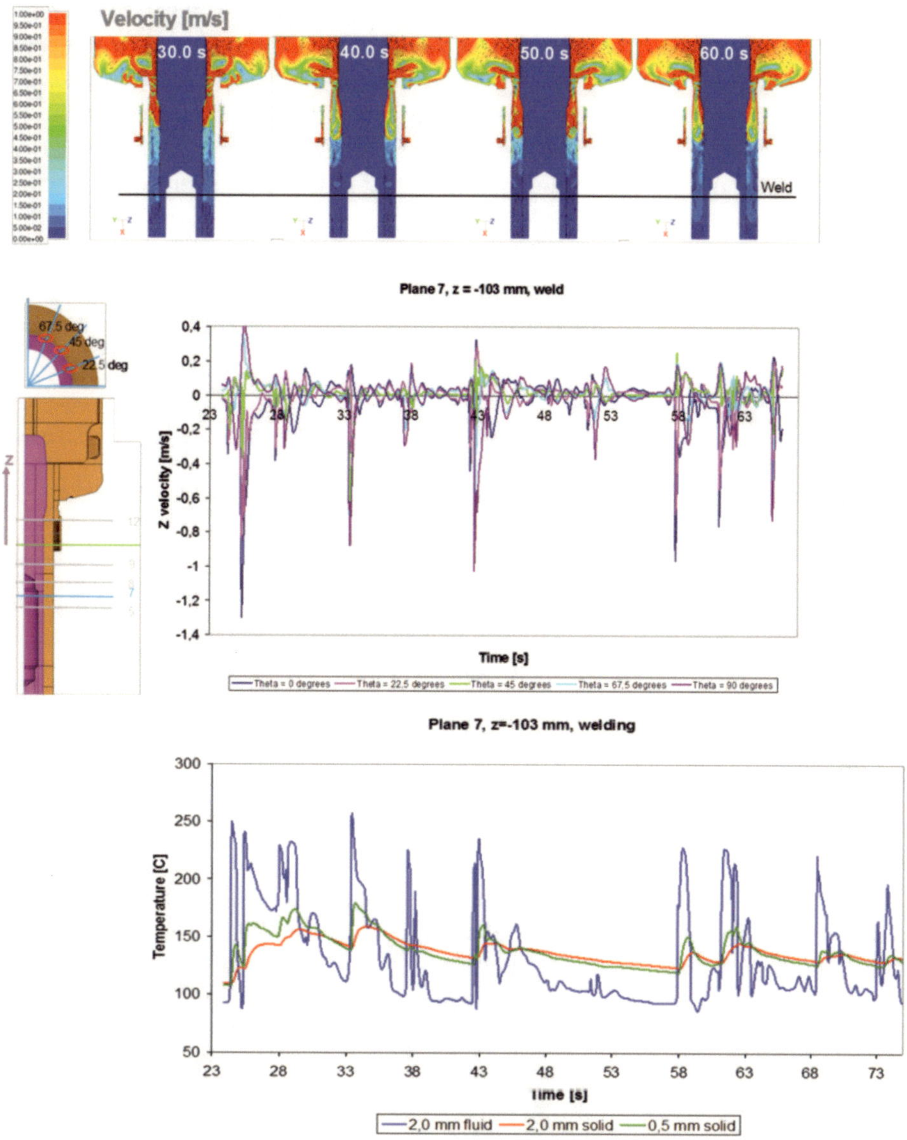

Figure 10. Upper view: velocity magnitude limited to 1 m/s (red ≥ 1 m/s). Middle view: time signal of axial vertical velocity component of the fluid in plane 7 (blue line in left reference picture) at 2 mm from the stem wall and for 5 different azimuthal positions. Lower view: temperature signal at three radial points at plane 7 with azimuthal angle of 45º.

However, the central question is still the quality of these simulations. Is a URANS simulation with good spatial and time resolutions enough to capture both the flow field and the heat transfer correctly? The author is tempted to answer with a "yes" but a more objective answer should be "probably". If the resolutions are adequate, the URANS approach based on SST k-ω model seems to work properly for cases with high enough Reynolds numbers and turbulent flow containing large perturbations. Furthermore, the comparison with experiments shown in Figure 11 exhibits a rather good agreement for mean temperature and temperature fluctuation distributions along a vertical line 1 mm apart from the stem wall. Incidentally, the material properties in the experiment are those of the prototype. Unfortunately, no comparisons of the velocity and velocity fluctuation fields were done but it is difficult to think that the agreement in the temperature fields shown in Figure 11 may be achieved with a radically different velocity field to that of the experiments.

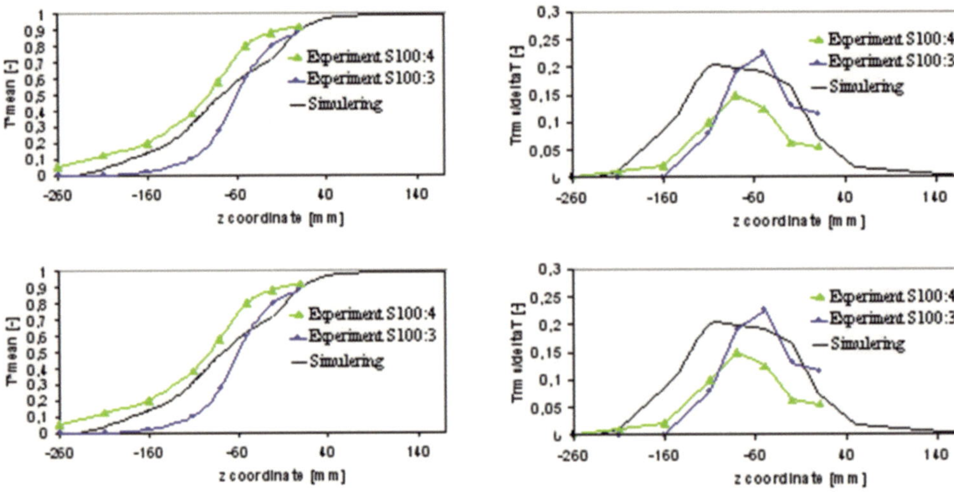

Figure 11. Comparison with the steel model experimental data (see Angele et al., 2011); left: mean temperature; right: temperature fluctuations (rms-value).

Then, are the simulated temperature fluctuations inside the solid stem also correct? This question is more difficult to answer since no experimental validation has been carried out but indirect indications about some properties of these fluctuations have nonetheless been obtained. Some of these properties are the order of magnitude of the most energy-containing frequencies of the temperature fluctuations at stem wall and of the associated heat transfer coefficients. Structural analyses (see e.g. Forsmarks Kraftgrupp, 2009) using this range of frequencies, together with the large temperature amplitudes and velocities of the order of 1 m/s, show that this kind of heat transfer is associated with the initiation of cracks on the stem surface. Moreover, the analyses also show that the time scale of the crack development that resulted in broken control rod stems is consistent both with the geometry of the stem (hollow welded joining; see Tinoco and Lindqvist, 2009) and the thermal loadings arising from the heat transfer process described by the CFD simulations. In general, the re-

http://dx.doi.org/10.5772/52691

sults of the structural analyses based on the aforementioned CFD-simulated conditions are consistent with the damage picture obtained by inspection. Using more "conservative" conditions, e.g. higher heat transfer coefficients, leads to unrealistic short time scales of the damage.

Furthermore, the mathematical and physical consistency of the temperature field inside the stem has been demonstrated by solving of the Inverse Heat Conduction Problem (IHCP), (Taler et al., 2011). The IHCP is defined as the determination of the boundary conditions of the problem from transient temperature histories given at one or more interior locations. The IHCP is a so called ill-posed problem, i.e. any small change on the input data can result in a dramatic change of the solution. By giving CFD-computed time signals of the temperature at selected points inside the stem, the temperature signal and the heat flux at the surface of the stem were recovered with good accuracy. The question that remains is the correctness of the temperature fluctuations at the surface of the stem, i.e. the correctness of the thermal interaction between the solid surface and the near wall region of the fluid. A definite answer can only be obtained by experimental validation or DNS.

Summing up, further structural analyses based on the same CFD-computed conditions have led to a set of measures that have been taken and/or are being implemented. The crud-cleaning flow has been modified to deliver a flow with a temperature of up to 100 ºC instead of the original of 60 ºC. After turning, a control rod stem may acquire surface tensile residual stress which increases the risk of crack initiation. In order to counteract this undesirable effect, all control rod stems are now burnished since this procedure creates surface compressive residual stress that increases the resistance of the stem against damage, i.e. increased protection against crack initiation and crack growth. At locations of the core most exposed to high warm bypass flows, a better material, stainless steel XM-19, has been chosen for the stems of the control rods. Only 50 of the 169 control rods will keep stems of a somewhat poorer material, stainless steel 316L. In the fuel outage of year 2014, all control rods will be tested in order to detect the presence of cracks and, based on the existent experience at that time, a renewed test program will be decided. This set of measures has been conceived as a viable alternative to the very expensive option of eliminating the cause of the mixing problem by changing the design of the control rod guide tubes comprising a new location of the upper bypass inlet.

6. Conclusion

This chapter constitutes a partial view of CFD, both in the sense of limited and favouring, since, for instance, the examples concerning BWR applications are overrepresented and important subjects such as nuclear safety and two-phase flows are left aside. As a kind of defence, it may be argued that the methods used for BWRs are valid, in general, for PWRs and, probably with important modifications, for SuperCritical Water Reactors (SCWRs). As far as CFD in nuclear safety is concerned, the high interest and large amount of work in the subject from, among others, the European Commission, (European Commission, 2005), and the U.S.

Department of Energy, Idaho National Laboratory, (Johnson et al., 2006), make and analysis and a review of the subject rather superfluous.

As pointed out by Tinoco et al. (2010c), two-phase flow simulations still have a relatively high degree of uncertainty and they have not reached the level of quality of single-phase simulations. The physics involved in two-phase phenomena is still not well understood, due mainly to the measuring difficulties entailed by the requirement of local knowledge of the processes needed in CFD. Consequently, the models available lack the distinctive prediction capability of CFD because they are usually based on information collected as relatively general correlations. A fundamental example is the experimental lack of knowledge about the modification by boiling of the detailed structure of a boundary layer at a wall. This deficiency leads to wrong predictions of flow resistance and void distribution in, for instance, pipe flow with boiling from the walls.

Another topic that has not been mentioned in this work is the field of acoustic perturbations in the reactor system, especially in the parts corresponding to pipe networks. The problem of flow-induced structural resonances has become more topical due to the increased interest in operating the reactors at uprated power output. The new components and/or the increased flow rates necessary for achieving the upgrade may lead to the generation of pressure waves that may excite structural vibration modes and increase the risk of mechanical fatigue. CFD may be used to analyse the pipe-network system if the geometry is realistically represented, the grid has a suitable resolution, the turbulence model is appropriate and, in the case of steam, compressibility is included (Lirvat et al., 2012). Post-processing of the generated data can provide information about the flow induced excitation modes, the source of the excitation frequencies/modes and possible standing pressure waves. In the past, attempts have been carried out to collect acoustic signals generated in complex systems like a nuclear reactor in order to detect possible deviations from normal operation of the different components. Any success has failed to turn up not due to the process of acquiring the data, which is relatively easy, but to the difficulties in interpreting the signal constituted by the superposition of broadcasted pressure waves from every component that may act as a source. In the future, CFD might help identifying the sources and describing how the different pressure waves propagate in the system. Perhaps, detecting the leakage of valves in complex systems like a nuclear reactor by means of acoustic signals may become possible.

As usual when trying to solve a problem, more questions than those initially asked arise during the solution process. The case of the control rod cracks in no exception and, as mentioned before, the apparent success of the URANS simulations with the SST k-ω model raises several questions, namely if this is the merit of the model, and at which Reynolds numbers the URANS approach begins to work, if the associate heat transfer, including temperature fluctuations, is correct, and with which resolutions in time and space, etc. All these questions should be investigated in the future since URANS may constitute a realistic alternative, especially at high Reynolds number, to the promising LES approach that still is too expensive for relatively large applications. Another alternative that has already been used with encouraging results, (Lindqvist, 2011), is Very Large Eddy Simulation (VLES) in the spirit of Speziale, (Speziale, 1998), an approach which has been further commented by Tinoco et al. (2010c).

Even if LES may be considered in many cases still too resource demanding, it is undeniably the approach of the near future provided that the computer development trend is maintained. However, conjugate heat transfer simulations of industrial applications, i.e. simulations at high Reynolds numbers, involved geometries and complex boundary conditions, appear to lack guidelines for ensuring acceptable quality with a suitable uncertainty. LES, as one of the simulation alternatives for these applications, seems to have taken a first step towards the definition of more complete rules for conjugate heat transfer simulations with the work of Carlsson and Veber (2010). Using a DNS data base for channel flow, (The University of Tokyo, 2012), mesh requirements for LES simulations of heat transfer are established, as functions of the Reynolds and Prandtl numbers. This work should be extended to conjugate heat transfer and temperature fluctuations in the solid by using the results of Tiselj et al. (2001), Tiselj et al. (2004), Tiselj (2011) and Tiselj and Cizelj (2012).

To conclude, this study has tried to show that the analysis of the mechanical integrity of light water reactors as a rather complex task, multidisciplinary and in continuous progress, evolving towards improved safety due to Regulatory demands, new and more advanced methods of analysis and taking advantage of the stably increasing available computer power. Additionally, this work has tried to describe, although incompletely and briefly, the historical progress of the currently accepted methodology and future prospects. Other types of analysis included in the existing methodology, which are not mentioned here due to space limitations, are, for example, studies of thermal loads in transients such as the loss of offsite power (Tinoco et al., 2005) or estimating the moderator tank maximum temperature due to gamma radiation and subcooled boiling heat transfer (Tinoco et al., 2008).

As it might have become apparent from this study, the FEM method for the analysis of structures is considered well established and sufficiently accurate, giving results that do not need verification or experimental validation. The case of three-dimensional numerical computations (CFD) applied to the thermal-hydraulics is very different, mainly due to turbulence, which still lacks an explanatory theory, and to the presence of two-phase flows where the knowledge about the physical processes involved is far from complete. Therefore, the results obtained using three-dimensional numerical computations require validation or at least verification, which is not always possible for complex systems such as nuclear reactors, an issue that has been discussed by Tinoco et al. (2010c).

Author details

Hernan Tinoco[1,2]

1 Forsmarks Kraftgrupp AB, Sweden

2 Onsala Ingenjörsbyrå AB, Sweden

References

[1] Ahlinder, S., Tinoco, H. and Ullberg, M., (2007). "CFD Analysis of Recombination by HWC in the Downcomer of a BWR", Proceedings of the 12th International Topical Meeting on Nuclear Reactor Thermal Hydraulics (NURETH-12), Pittsburgh, Pennsylvania, U.S.A., September 30-October 4, 2007.

[2] Angele, K., Odemark, Y., Cehlin, M., Hemström, B., Högström, C.-M., Henriksson, M., Tinoco, H. and Lindqvist, H., (2011). "Flow Mixing Inside a Control-Rod Guide Tube – Part II: Experimental Tests and CFD", Nuclear Eng. Design, Vol. 241, Issue 12, December, 2011.

[3] ANSI/ANS, (1983a). "Nuclear Safety Criteria for the Design of Stationary Pressurized Water Reactor Plants", ANSI/ANS-51.1-1983.

[4] ANSI/ANS, (1983b). "Nuclear Safety Criteria for the Design of Stationary Boiling Water Reactor Plants", ANSI/ANS-52.1-1983.

[5] ANSI/ANS, (1988). "Design Basis for Protection of Light Water Nuclear Power Plants Against the Effects of Postulated Pipe Rupture", ANSI/ANS-58.2-1988.

[6] ASME, (2010). "Boiler and Pressure Vessel Code with ADDENDA, III: Rules for Construction of Nuclear Power Plant Components", 2010 Edition, New York, USA.

[7] Brodkey, R. S., (1967). The Phenomena of Fluid Motions, Dover, New York, USA.

[8] Carlsson, F. and Veber, P., (2010). "Heat transfer validation using LES in turbulent Channel flow", Nordic Symposium Kärnteknik, Nuclear Technology 2010, December 2, Stockholm, Sweden.

[9] Casey, M. and Wintergerste, T., Editors, (2000). "Best Practice Guidelines", ERCOFTAC Special Interest Group on "Quality and Trust in Industrial CFD", SulzerInnotec, Version 1.0, Switzerland.

[10] Châtelain, A., Ducros, F. and Métais, O., (2003). "LES of conjugate heat-transfer using thermal wall-functions", Proceedings of the D-LES 5 ERCOFTAC Workshop, Munich, Germany, pp. 307-314.

[11] Craft, T. J., Iacovides, H. and Uapipatanakul, S., (2010). "Towards the development of RANS models for conjugate heat transfer", J. Turbulence, Vol. 11, No. 26, pp. 1-16.

[12] de Tilly, A., Penot, F. and Tuhault, J. L. (2006). "Instantaneous heat transfer measurements between an unsteady fluid flow and a wall: application to a non-isothermal 2D-mixing tee", Int. Heat Transfer Conf., Australia, IHTC13. p21.80.

[13] de Tilly, A. and Sousa, J. M. M., (2008a). "An experimental study of heat transfer in a two-dimensional T-junction operating at a low momentum flux ratio", Int. J. Heat Mass Transfer, Vol. 51, pp. 941–7.

[14] de Tilly, A. and Sousa, J. M. M., (2008b). "Measurements of instantaneous heat transfer in a two-dimensional T-junction using a multi-thermocouple probe", Meas. Sci. Technol., Vol. 20, pp. 045106-1 – 045106-9.

[15] Dimotakis, P. E., (2005). "Turbulent Mixing", Annual Rev. Fluid Mech. 37:329-56.

[16] Ellenberger, J. P., Chuse, R., Carson, B., (2004). "Pressure Vessels the ASME Code Simplified", Eighth Edition, McGraw-Hill, New York, USA.

[17] EU, (1997). "The Pressure Equipment Directive (PED)" (97/23/EC), EU home page, http://ec.europa.eu/enterprise/sectors/pressure-and-gas/documents/ped/index_en.htm.

[18] European Commission, (2005). "ECORA - Evaluation of Computational Fluid Dynamics Methods for Reactor Safety Analysis", Condensed Final Summary Report, https://domino.grs.de/ecora/ecora.nsf/ .

[19] ForsmarksKraftgrupp, (2007). "Forsmark 1 – KFM - Design Specifications for pressure- and force-bearing components of system 415 – Feedwater System", Forsmark Report F1-KFM-415 (in Swedish), 12 June, 2007.

[20] Forsmarks Kraftgrupp, (2009). "Forsmark 3 – System 30-222, Control Rods. Restart of Forsmark 3 after the refuelling outage of 2009 – Analysis with respect to mechanical integrity", Report FT-2009-2723 (in Swedish), Forsmark, Sweden.

[21] Hemström, B., Lundström, A., Tinoco, H. and Ullberg, M., (1992). "Flow field dependence of reactor water chemistry in BWR", Proceedings of the 6th International Conference on Water Chemistry of Nuclear Reactor Systems, British Nuclear Energy Society, Bournemouth, 12-15 October, 1992.

[22] Hermansson, M. and Thorsson, A., (1997). "Forsmark 1 and 2 – Core grid. System 212.3. Dynamic Analysis of Reactor Pressure Vessel with Internals. Response Spectra and forces in core grid for load cases SSE, SRV-1, SRV-12, CO, Chugging and PBI." Forsmark report FT-1997-442, rev 3.

[23] Hill, R. S., (2008). "Section III – Component Design and Construction", Supporting New Build and Nuclear Manufacturing in South Africa, ASME Nuclear Codes and Standards Workshop, Sandton, South Africa, October 7 – 8, 2008.

[24] Hosseini, S. M., Yuki, H., and Hashizume, H., (2008). "Classification of Turbulent Jets in a T-Junction Area with a 90-deg Bend Upstream", Int. J. Heat Mass Transfer, Vol. 51 (9-10), pp. 2444–2454.

[25] Hosseini, S. M., Yuki, H., and Hashizume, H., (2009). "Experimental Investigation of Flow Field Structure in Mixing Tee" J. Fluids Eng., Vol. 131, pp. 051103-1 - 051103-7.

[26] IAEA, (2010), "Safety Classification of Structures, Systems and Components in Nuclear Power Plants", IAEA Safety Standards, DS367.

[27] Jayaraju, S. T., Komen, E. M. J. and Baglietto, E., (2010). "Suitability of wall-functions in Large Eddy Simulations for thermal fatigue in a –junction", Nuc. Eng. Design, Vol. 240, pp. 2544-2554.

[28] Johnson, R. W., Schultz, R. R., Roache, P. J., Celik, I. B., Pointer, W. D. and Hassan, Y. A., (2006). "Processes and Procedures for Application of CFD to Nuclear Reactor Safety Analysis", Idaho National Laboratory, Report INL/EXT-06-11789, USA, September 2006.

[29] Kasagi, N., Kuroda, A., and Hirata, M., (1989). ''Numerical Investigation of Near-Wall Turbulent Heat Transfer Taking Into Account the Unsteady Heat Conduction in the Solid Wall,'' J. Heat Transfer, Vol. 111, pp. 385–392.

[30] Keshmiri, A., (2006). "Modelling of Conjugate Heat Transfer in Near-Wall Turbulence", MSc Thesis, University of Manchester, UK.

[31] Kuhn, S., Braillard, O., Ničeno, B. and Prasser, H.-M., (2010). "Computational study of conjugate heat transfer in T-junctions", Nuc. Eng. Design, Vol. 240, pp. 1548-1557.

[32] Kukukova, A., Aubin, J. and Kresta, S., (2009). "A new definition of mixing and segregation: Three dimensions of a key process variable", Chem. Eng. Res. Design, Vol. 87, No. 4, pp. 633-647.

[33] Lahey, R. T. and Moody, F. J., (1993). "The Thermal-Hydraulics of a Boiling Water Nuclear Reactor", ANS, La Grange Park, Illinois, USA.

[34] Lindqvist, H., (2011). "CFD Simulation on Flow Induced Vibrations in High Pressure Control and Emergency Stop Turbine Valve", 14th International Topical Meeting on Nuclear Reactor Thermalhydraulics, NURETH-14, Toronto, Ontario, Canada, September 25-30, 2011.

[35] Lirvat, J., Mendonça, F., Veber, P. and Andersson, L., (2012). "Flow and Acoustic Excitation mechanisms in a Nuclear Reactor Steam-line and Branch Network", 18th AIAA/CEAS Aeroacoustics Conference, Colorado Springs, Colorado, USA, 4–6 June 2012.

[36] Menter, F., (2002). "CFD Best Practice Guidelines for CFD Code Validation for Reactor-Safety Applications", UE/FP5 ECORA Project "Evaluation of computational fluid dynamic methods for reactor safety analysis", EVOL-ECORA-D01, Germany.

[37] Mosyak, A., Pogrebnyak, E., and Hetsroni, G., (2001). ''Effect of Constant Heat Flux Boundary Condition on Wall Temperature Fluctuations,'' ASME J. Heat Transfer, Vol. 123, pp. 213–218.

[38] Muramatsu, T., (2003). "Generation Possibilities of Lower Frequency Components in Fluid Temperature Fluctuations Related to Thermal Striping Phenomena", Transactions of the 17th International Conference on Structural Mechanics in Reactor Technology (SMIRT 17), Prague, Czech Republic, August 17-22, 2003.

[39] Naik-Nimbalkar, V. S., Patwardhan, A. W., Banerjee, I., Padmakumar, G. and Vaidyanathan, G., (2010). "Thermal Mixing in T-junctions", Chem. Eng. Science, Vol. 65, pp. 5901-5911.

[40] OECD/NEA, (2011). "Report of the OECD-NEA-Vattenfall T-Junction Benchmark exercise", Report NEA/CSNI/R(2011)5, 12 May, 2011.

[41] Ogawa, H., Igarashi, M., Kimura, N. and Kamide, H., (2005)."Experimental Study on Fluid Mixing Phenomena in T-pipe Junction with Upstream Elbow", Proceedings of the 11th International Topical Meeting on Nuclear Reactor Thermal Hydraulics (NURETH 11), Popes Palace Conference Center, Avignon, France, October 2-6, 2005.

[42] Pasutto, T., Péniguel, C. and Sakiz, M., (2006)."Chained Computations Using an unsteady 3D Approach for the Determination of Thermal Fatigue in a T-Junction of a PWR Nuclear Plant", Nuc. Eng. Tech., Vol. 38, No. 2, pp. 147-154.

[43] Polyakov, A. F., (1974). "Wall Effect of Temperature Fluctuations in the Viscous Sublayer", TeplofizikaVysokihTemperatur, Vol. 12, pp. 328–337.

[44] Pozorski, J. and Minier, J.-P., (2005). "Stochastic Modelling of Conjugate Heat Transfer in Near-Wall Turbulence", Engineering Turbulence Modelling and Experiments 6, W. Rodi Editor, Elsevier, pp. 803-812.

[45] Speziale, C., (1998). "Turbulence Modeling for Time-Dependent RANS and VLES: A Review", AIAA Journal, Vol. 36, No.2, pp.173-184.

[46] STUK, (2000). "Nuclear Power Plant Systems, Structures and Components and their Safety Classification", Guide YVL 2.1, 26 June, 2000.

[47] Taler, J., Cebula, A., Marcinkiewicz, J. and Tinoco, H., (2011). "Heat Flux and Temperature Determination on the Control Rod Outer Surface", 14th International Topical Meeting on Nuclear Reactor Thermalhydraulics, NURETH-14, Toronto, Ontario, Canada, September 25-30, 2011.

[48] The Marviken Full Scale Critical Flow Tests, (1979). "Summary Report, Joint Reactor Safety Experiments in the Marviken Power Station", StudsvikEnergiteknik A.B., Nyköping, December 1979, Sweden.

[49] The University of Tokyo, (2012). Department of Mechanical Engineering, Turbulence and Heat Transfer Laboratory, http://www.thtlab.t.u-tokyo.ac.jp/.

[50] Tinoco, H., (2001). "A Standing Pressure Wave Hypothesis of Oscillating Forces Generated During a Steam-Line Break", proceedings of the 9th International Conference on Nuclear Engineering (ICONE 9), Nice, France, April 8-12, 2001.

[51] Tinoco, H., (2002). "Three-Dimensional Modeling of a Steam-Line Break in a Boiling Water Reactor", Nuclear Sci. and Eng., Vol. 140, pp. 152-164.

[52] Tinoco, H., Adolfsson, E., Baltyn, W. and Marcinkiewicz, J., (2005). "BWR ECCS – Thermal Loading Analysis of Internals Related to Spray System Removal", proceed-

ings of the 13th International Conference on Nuclear Engineering (ICONE 13), Beijing, China, May 16-20, 2005.

[53] Tinoco, H., Ahlinder, S. and Hedberg, P., (2008). "Estimate of Core Shroud Temperature by means of a CFD-Model of the Core Bypass", proceedings of the 15th International Conference on Nuclear Engineering (ICONE 15), Nagoya, Japan, April 22-26 (2007) and Journal of Power and Energy Systems, Vol. 2, No. 2, pp. 775-789, 2008.

[54] Tinoco, H. and Lindqvist, H., (2009). "Thermal Mixing Instability of the Flow Inside a Control-Rod Guide Tube", proceedings of the 13th International Topical Meeting on Nuclear Reactor Thermal Hydraulics (NURETH 13), Kanazawa City, Ishikawa Prefecture, Japan, September 27-October 2, 2009.

[55] Tinoco, H. and Ahlinder, S., (2009). "Mixing Conditions in the Lower Plenum and Core Inlet of a BWR", J. Eng. Gas Turbines & Power, Vol. 131, 062903-1 - 062903-12, November, 2009.

[56] Tinoco, H., Buchwald, P. and Frid, W., (2010a). "Numerical Simulation of Boron Injection in a BWR", Nuclear Eng. Design, Vol. 240, Issue 2, February, 2010.

[57] Tinoco, H., Lindqvist, H., Odemark, Y., Högström, C.-M., Hemström, B. and Angele, K., (2010b). "Flow Mixing Inside a Control-Rod Guide Tube – Part I: CFD Simulations", Proceedings of the 18th International Conference on Nuclear Engineering (ICONE 18), Xi'an, China, May 17-21, 2010.

[58] Tinoco, H., Lindqvist, H. and Frid, W., (2010c). "Numerical Simulation of Industrial Flows", Numerical Simulations - Examples and Applications in Computational Fluid Dynamics, Lutz Angermann (Ed.), ISBN: 978-953-307-153-4, InTech, Available from: http://www.intechopen.com/books/numerical-simulations-examples-and-applications-in-computational-fluid-dynamics/numerical-simulation-of-industrial-flows.

[59] Tiselj, I., Bergant, R., Mavko, B., Bajsić, I. and Hetsroni, G., (2001). "DNS of turbulent heat transfer in channel flow with heat conduction in the wall", J. Heat Transfer – Transactions of ASME, Vol. 123 (5), pp. 849-857.

[60] Tiselj, I., Horvat, A., Mavko, B., Pogrebnyak, E., Mosyak, A. and Hetsroni, G., (2004). "Wall properties and heat transfer in near wall turbulent flow", Numerical Heat Transfer A, Vol. 46, pp. 717-729.

[61] Tiselj, I. (2011). "DNS of Turbulent Channel Flow at Re_T = 395, 590 and Pr = 0.01", 14th International Topical Meeting on Nuclear Reactor Thermalhydraulics (NURETH-14), Toronto, Ontario, Canada, September 25-30, 2011.

[62] Tiselj, I. and Cizelj, L., (2012). "DNS of turbulent channel flow with conjugate heat transfer at Prandtl number 0.01", Nuc. Eng. Design - in print.

[63] USGenWeb, (2011). "The R. B. Grover & Company Shoe Factory Boiler Explosion", http://plymouthcolony.net/brockton/boiler.html.

[64] vanHaren, S. W. (2011). "Testing DNS capability of OpenFOAM and STAR-CCM+", Master Thesis, University of Delft, September 15, 2011.

[65] Varrasi, J., (2005). "The True Harnessing of Steam", Mechanical Eng. Mag., January 2005, New York, USA.

[66] Westin, J., 't Mannetje, C., Alavyoon, F., Veber, P., Andersson, L., Andersson, U., Eriksson, J., Henriksson, M., & Andersson, C., (2008). "High-cycle thermal fatigue in mixing tees. Large-eddy simulations compared to a new validation experiment", proceedings of the 16th International Conference on Nuclear Engineering (ICONE 18), Orlando, Florida, USA, May 11-15, 2008.

[67] Zeng, L., Jansson, L.G. and Dahlström, L., (2011). "Non-Linear Design Evaluation of Class 1-3 Nuclear Power Piping", Nuclear Power - System Simulations and Operation, PavelTsvetkov (Ed.), ISBN: 978-953-307-506-8, InTech, Available from: http://www.intechopen.com/books/nuclear-power-system-simulations-and-operation/non-linear-design-evaluation-of-class-1-3-nuclear-power-piping.

Section 2

General Thermal Hydraulic Applications

Chapter 4

Thermal Hydraulics Design and Analysis Methodology for a Solid-Core Nuclear Thermal Rocket Engine Thrust Chamber

Ten-See Wang, Francisco Canabal, Yen-Sen Chen,
Gary Cheng and Yasushi Ito

Additional information is available at the end of the chapter

http://dx.doi.org/10.5772/52569

1. Introduction

Nuclear thermal propulsion [1] can carry far larger payloads and reduce travel time for astronauts going to Mars than is now possible with chemical propulsion. The most feasible concept, extensively tested during the Rover/NERVA (Nuclear Engine for Rocket Vehicle Application) era, is the solid-core concept [2]. It features a solid-core nuclear reactor consisting of hundreds of heat generating prismatic flow elements. Each flow element contains tens of tubular flow channels through which the working fluid, hydrogen, acquires energy and expands in a high expansion nozzle to generate thrust. Approximately twenty nuclear thermal engines with different sizes and designs were tested during that era. Using hydrogen as propellant, a solid-core design typically delivers specific impulses (ISP) on the order of 850 to 1000 s, about twice that of a chemical rocket such as the Space Shuttle Main Engine (SSME).

With the announcement of the Vision for Space Exploration on January 14, 2004, NASA Marshall Space Flight Center started a two-year solid-core nuclear rocket development effort in 2006. The tasks included, but not limited to, nuclear systems development, design methodology development, and materials development. In 2011, with the retirement of Space Shuttle fleets, and NASA's shifting focus to further out places such as Mars and asteroids, nuclear thermal propulsion is likely to garner substantial interest again. It is therefore timely to discuss the design methodology development effort from 2006 to 2007, entitled "Multiphysics Thrust Chamber Modeling", which developed an advanced thermal hydraulics computational methodology and studied a solid-core, nuclear thermal engine designed in the Rover/NREVA era.

One of the impacts made by this thermal hydraulics design methodology is the consideration of chemical reacting flow that addresses the effect of hydrogen decomposition and recombination. The advantage of using hydrogen as a propellant is well known in the chemical rocket due to its low molecular weight. However, molecular hydrogen decomposes to atomic hydrogen during high-temperature heating in the thermal nuclear reactor. Since atomic hydrogen has half the weight of that of molecular hydrogen, it was speculated by some that the total thrust could be doubled if all of the hydrogen is dissociated at very high temperatures, therefore leaning to the high power density reactor design. In actuality, the hydrogen conversion is often not uniform across the solid-core since the reactor temperature depends on the hydrogen flow distribution and the actual power profile generated by the nuclear fuel. In addition, the hydrogen atoms recombine in the nozzle since temperature decreases rapidly during the gas expansion, thereby negating the thrust gain. To the best of our knowledge, however, the effect of hydrogen decomposition in a thermal nuclear thrust chamber on the thrust performance has never been studied.

On the other hand, it is always desirable to decrease the reactor weight while one of the ideas is to reduce the reactor size, which increases the power density. One of the impacts of operating at the combination of high temperature and high power density is a phenomenon known as the mid-section corrosion [3], as reported during the legacy engine tests. It is the cracking of the coating layer deposited on the inner wall of the flow channel, coupled with an excessive mass loss of the material near the mid-section of a flow element. The purpose of the coating layer was to isolate the carbonaceous compound in the flow element matrix from the attack by hydrogen. The causes of mid-section corrosion were speculated as a mismatch in the thermal expansion of flow element and its coating material, high flow element web internal temperature gradients, and change of solid thermal property due to irradiation [3, 4]. Those are all possibilities related to the materials. We, however, wanted to trace the cause from a thermal-hydraulics design view point. That is, with the long, narrow flow channel design that is used to heat the hydrogen, the possibility of flow choking in the channel was never studied. One of the efforts in this task was therefore to investigate the possibility of chocked flow occurring in the long, narrow flow channel, and its implication on heat accumulation in the flow element and mid-section corrosion.

The objective of this study was to bridge the development of current design methods with those developed in the Rover/NERVA era, thereby providing a solid base for future development. This is because during the Rover/NERVA era, there was a wealth of lessons learned from those legacy engine tests. All those lessons learned culminated in the final design of the Small Engine [5], but was never built nor tested since it was designed near the end of that era. The Small Engine was therefore a 'paper engine' that bears the best features of lessons learned. By simulating and comparing the computed environments with those of the Small Engine design analysis and available test information from the legacy tests, the lessons learned during Rover/NERVA era may be revitalized and the effect of some important design features may be validated. This Chapter therefore reviews the thermal hydraulics computational methodology developed to help the design of a materials tester [4, 6], and the results reported in the design analyses of the Small Engine [7, 8]. Figure 1 shows the computational model [7] of the Small Engine that shows the surfaces of the hydrogen inlet duct, a pressure vessel that houses the solid-core reactor, and an exhaust nozzle that provides the thrust.

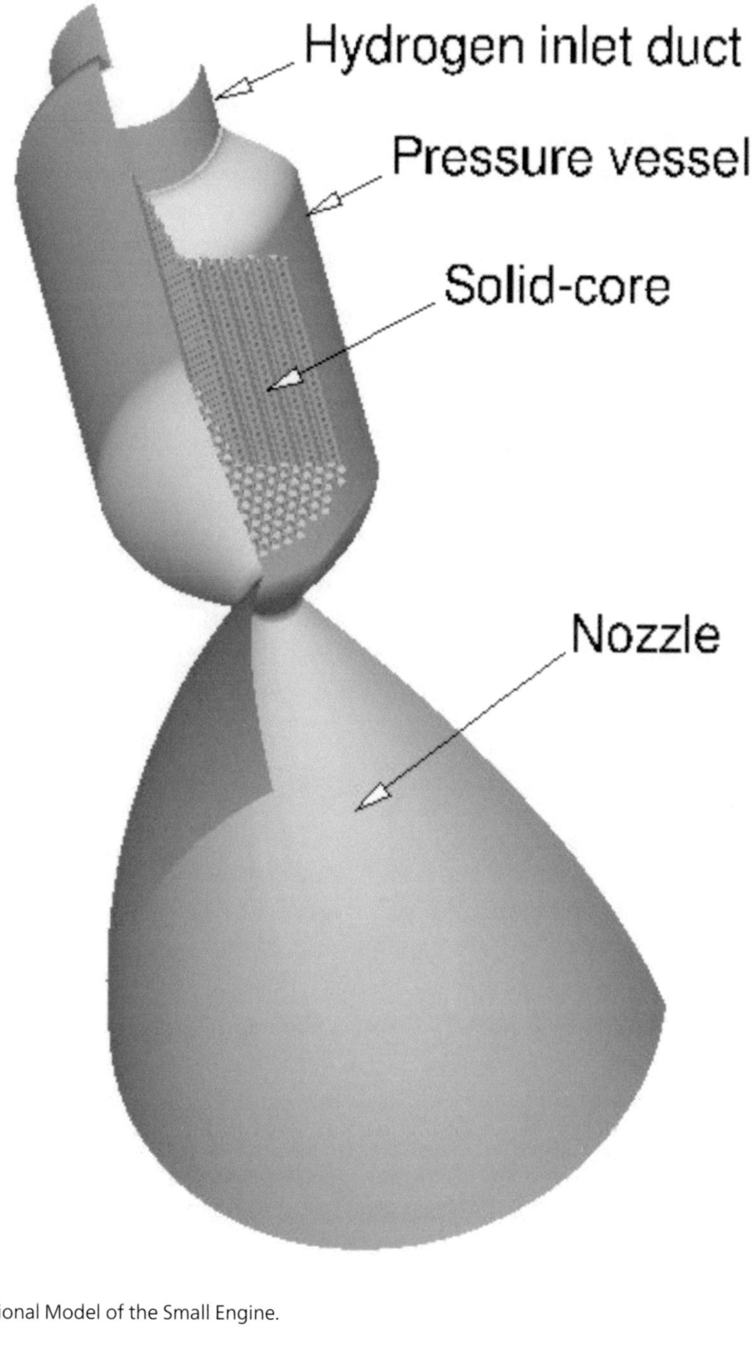

Figure 1. Computational Model of the Small Engine.

A two-pronged approach was conducted to address the aforementioned efforts: a detailed modeling of a single, powered flow element that addresses possible, additional cause of the mid-section corrosion [8], and a global modeling approach that computes the entire thermal flowfield of the Small Engine thrust chamber [7]. The latter links the thrust performance with the effects of power profiles, chemical reactions, and overall heat transfer efficiency. The global approach solves the entire thrust chamber with a detailed modeling, except the thousands of flow channels in the hundreds of flow elements were lumped together as a porous media, for computational efficiency. The heat transfer between other supporting solid components and the working fluid is solved with the conjugate transfer methodology, which was developed in [4, 6]. Theoretical and neutronics provided power profiles were used in lieu of the coupled neutronics modeling. The computational methodology, the results of the simulations of a single flow element and that of the entire thrust chamber, are presented and discussed herein.

2. Computational heat transfer methodology

2.1. Fluid dynamics

The computational methodology was based on a multi-dimensional, finite-volume, viscous, chemically reacting, unstructured grid, and pressure-based, computational fluid dynamics formulation [9]. Time-varying transport equations of continuity, species continuity, momentum, total enthalpy, turbulent kinetic energy, and turbulent kinetic energy dissipation rate were solved using a time-marching sub-iteration scheme and are written as:

$$\frac{\partial \rho}{\partial t}+\frac{\partial}{\partial x_j}\left(\rho u_j\right)=0 \tag{1}$$

$$\frac{\partial \rho \alpha_i}{\partial t}+\frac{\partial}{\partial x_j}\left(\rho u_j \alpha_i\right)=\frac{\partial}{\partial x_j}\left[\left(\rho D+\frac{\mu_t}{\sigma_\alpha}\right)\frac{\partial \alpha_i}{\partial x_j}\right]+\omega_i \tag{2}$$

$$\frac{\partial \rho u_i}{\partial t}+\frac{\partial}{\partial x_j}\left(\rho u_j u_i\right)=-\frac{\partial p}{\partial x_i}+\frac{\partial \tau_{ij}}{\partial x_j} \tag{3}$$

$$\begin{aligned}\frac{\partial \rho H}{\partial t}+\frac{\partial}{\partial x_j}\left(\rho u_j H\right)=\frac{\partial p}{\partial t}+\frac{\partial}{\partial x_j}\left[\left(\frac{K}{C_p}+\frac{\mu_t}{\sigma_H}\right)\frac{\partial H}{\partial x_j}\right]+\frac{\partial}{\partial x_j}\left[\left(\left(\mu+\mu_t\right)-\left(\frac{K}{C_p}+\frac{\mu_t}{\sigma_H}\right)\right)\frac{\partial\left(V^2/2\right)}{\partial x_j}\right]\\+\frac{\partial}{\partial x_j}\left[\left(\frac{K}{C_p}+\frac{\mu_t}{\sigma_H}\right)\left(u_k\frac{\partial u_j}{\partial x_k}-\frac{2}{3}u_j\frac{\partial u_k}{\partial x_k}\right)\right]+Q_r\end{aligned} \tag{4}$$

$$\frac{\partial pk}{\partial t}+\frac{\partial}{\partial x_j}(pu_j\left[\left(\mu\frac{\mu_t}{\delta_k}\right)\frac{\partial k}{\partial x_j}\right]+p(\Pi-\varepsilon) \tag{5}$$

$$\frac{\partial p\varepsilon}{\partial t}+\frac{\partial}{\partial x_j}\left(pu_{j\varepsilon}\right)=\frac{\partial}{\partial x_j}\left[\left(\mu+\frac{\mu_t}{\delta_\varepsilon}\right)\frac{\partial \varepsilon}{\partial x_j}\right]+p\frac{\varepsilon}{k}\left(C_j\Pi-C_2\varepsilon+C_3\Pi^2/\varepsilon\right) \tag{6}$$

A predictor and corrector solution algorithm was employed to provide coupling of the governing equations. A second-order central-difference scheme was employed to discretize the diffusion fluxes and source terms. For the convective terms, a second-order upwind total variation diminishing difference scheme was used. To enhance the temporal accuracy, a second-order backward difference scheme was employed to discretize the temporal terms. Details of the numerical algorithm can be found in [8-13].

An extended k-ε turbulence model [14] was used to describe the turbulent flow and turbulent heat transfer. A modified wall function approach was employed to provide wall boundary layer solutions that are less sensitive to the near-wall grid spacing. Consequently, the model has combined the advantages of both the integrated-to-the-wall approach and the conventional law-of-the-wall approach by incorporating a complete velocity profile and a universal temperature profile [10].

2.2. Heat transfer in solids

A solid heat conduction equation was solved with the gas-side heat flux distributions as its boundary conditions. The solid heat conduction equation can be written as:

$$\frac{\partial pC_pT_s}{\partial t}=\frac{\partial}{\partial x_j}\left(K\frac{\partial T_s}{\partial x_i}\right)+Q_p+Q_s \tag{7}$$

The present conjugate heat transfer model [4] solves the heat conduction equation for the solid blocks separately from the fluid equations. The interface temperature between gas and solid, which is stored at interior boundary points, is calculated using the heat flux continuity condition. For solution stability and consistency, the gas and solid interface boundary temperature is updated using the transient heat conduction equation (7).

2.3. Flow and heat transfer in porous media

A two-temperature porosity model was formulated with separate thermal conductivities for the flow and the solid parts. The heat transfer between the flow and solid was modeled by using the empirical correlation of the heat transfer coefficient for circular pipes as a function of flow Reynolds numbers. Empirical multipliers for both the heat transfer and drag loss were determined by comparing solutions of flow passing through a porous flow element with those of a Small Engine 19-channel flow element using detailed conjugate heat transfer modeling [8]. The only affected fluid governing equations are Navier-Stokes and energy equations and can be rewritten as:

$$\frac{\partial \rho u_i}{\partial t}+\frac{\partial}{\partial x_j}\left(\rho u_j u_i\right)=-\frac{\partial p}{\partial x_i}+\frac{\partial \tau_{ij}}{\partial x_j}-\frac{L}{\beta} \tag{8}$$

$$\begin{aligned}\frac{\partial \rho H}{\partial t}+\frac{\partial}{\partial x_j}\left(\rho u_j H\right)=\frac{\partial p}{\partial t}+\frac{\partial}{\partial x_j}\left[\left(\frac{K}{C_p}+\frac{\mu_t}{\sigma_H}\right)\frac{\partial H}{\partial x_j}\right]+\frac{\partial}{\partial x_j}\left[\left((\mu+\mu_t)-\left(\frac{K}{C_p}+\frac{\mu_t}{\sigma_H}\right)\right)\frac{\partial\left(V^2/2\right)}{\partial x_j}\right] \\ +\frac{\partial}{\partial x_j}\left[\left(\frac{K}{C_p}+\frac{\mu_t}{\sigma_H}\right)\left(u_k\frac{\partial u_j}{\partial x_k}-\frac{2}{3}u_j\frac{\partial u_k}{\partial x_k}\right)\right]+Q_r+\frac{1}{\beta}Q_s\end{aligned} \tag{9}$$

For the solid heat conduction in porous media,

$$\frac{\partial \rho_s C_{ps} T_s}{\partial t}=\frac{\partial}{\partial x_j}\left(K_s\frac{\partial T_s}{\partial x_i}\right)+\frac{1}{1-\beta}\left(Q_p-Q_s\right) \tag{10}$$

For the Small Engine 19-channel flow element heat-exchanger configuration, drag loss for flow in circular pipes can be used as a point of departure. That is,

$$L=\frac{1}{2}\rho f_L c_f |U| u_i / d \tag{11}$$

where $c_f = 0.0791 Re^{-0.25}$ is the Blasius formula for turbulent pipe flow [15]. Typical Reynolds numbers in a flow channel range from 10,000 to 40,000.

For the heat exchange source term,

$$Q_s=\frac{1}{2}\rho f_q \frac{c_f}{P_r^{2/3}}|U| C_p (T_s - T)/d \tag{12}$$

For the purpose of this study, the conjugate heat transfer module for solids was benchmarked with the analysis of a cylindrical specimen heated by an impinging hot hydrogen jet [4]. The computed solid temperature profiles agreed well with those of a standard solid heat transfer code SINDA [16]. The methodology for flow through porous media was verified through a particle-bed nuclear flow element [17] and the Space Shuttle Main Engine (SSME) main injector assembly [18]. The numerical and physical models for predicting the thrust and wall heat transfer were benchmarked with an analysis of the SSME thrust chamber flowfield, in which the computed axial-thrust performance, flow features, and wall heat fluxes compared well with those of the test data [12].

3. Small engine

The goals of this study were achieved by computing the thermal hydraulics of a flow element and the entire thrust chamber of an engine designed near the end of the Rover/NERVA era – the Small Engine. The thrust chamber of Small Engine composes of an inlet plenum, the solid-core nuclear reactor or heat exchanger, and an exhaust nozzle, as shown in Fig. 1. There are 564 flow elements and 241 support elements or tie-tubes designed for the thermal nuclear reactor. The flow element is shaped like a hexagonal prism, with a length of about 890 mm and a width of about 19 mm from flat to flat [5, 8]. The prismatic flow element contains 19 tubular coolant channels. Three coolant channel diameters were designed for the Small Engine. Each flow element is held in position by three tie-tubes and the corresponding hot-end support system (not modeled). General geometry and operating conditions were obtained from [5], while certain specific operating conditions and nozzle geometry were calculated and provided by the Systems Engineering group.

3.1. Power profiles

For the purpose of this study, theoretical and neutronics calculation provided power profiles were imposed onto the solid-core domain in lieu of the coupled neutronics calculations, for computational efficiency. Combinations of two axial and three radial power profiles, as shown in Figs. 2 and 3, were used to show the effect of which on the heat transfer and thrust performance. Figure 2 shows a Cosine profile and a clipped Cosine profile for the power distribution in the axial direction of the solid-core reactor, while Fig. 3 shows a Cosine profile, a flattened (Cosine) profile, and a flat profile for the power distributions in the radial direction of the nuclear heat exchanger. Given an example for the thrust chamber computations, three combinations of these power profiles were assumed. The first combination uses the shape of the Cosine curve (shown in Figs. 2 and 3) in both the axial and radial directions. That combined power distribution resembles the thermal flux distribution in bare reactors [19], and is simply named as the Cosine-Cosine power profile. By definition, the combined Cosine-Cosine power profile peaks at the middle of the core and drops to zero at the core boundary due to escaping neutrons. The second combination was prescribed by a neutronics calculation with varied Uranium loading. It features the clipped Cosine profile (shown in Fig. 2) in the axial direction and the flattened (Cosine) profile (shown in Fig. 3) in the radial direction, and is dubbed as the clipped Cosine-flattened power profile. The varied fuel loading flattens the (Cosine) power profile in the radial direction, but the power rises near the boundary to show the effect of the reflector, as shown in Fig. 3. The idea of flattening the radial profile is such that the flow in the channels is heated more uniformly, thereby improving the heat transfer efficiency. It is envisioned that the clipped Cosine-flattened power profile is probably the closest power profile to that intended for the Small Engine.

It can be seen that if a radially flattened power profile improves the heat transfer efficiency, then a theoretically flat, radial power profile should reach even higher heat transfer efficiency. We therefore proposed a flat power profile design for the radial direction. A theoretically flat radial power distribution may be achieved with a combination of varied fuel loading and working fluid flow distribution. The third combination therefore employs the clipped Cosine curve (shown in Fig. 2) in the axial direction and a flat curve (shown in Fig. 3) in the radial direction, and is called the clipped Cosine-flat power profile.

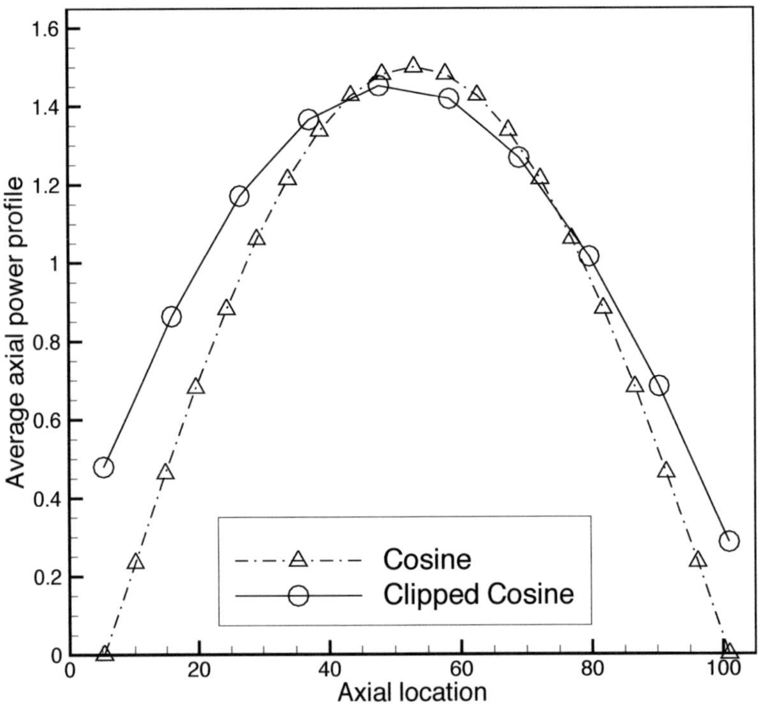

Figure 2. Power profiles used in the axial direction

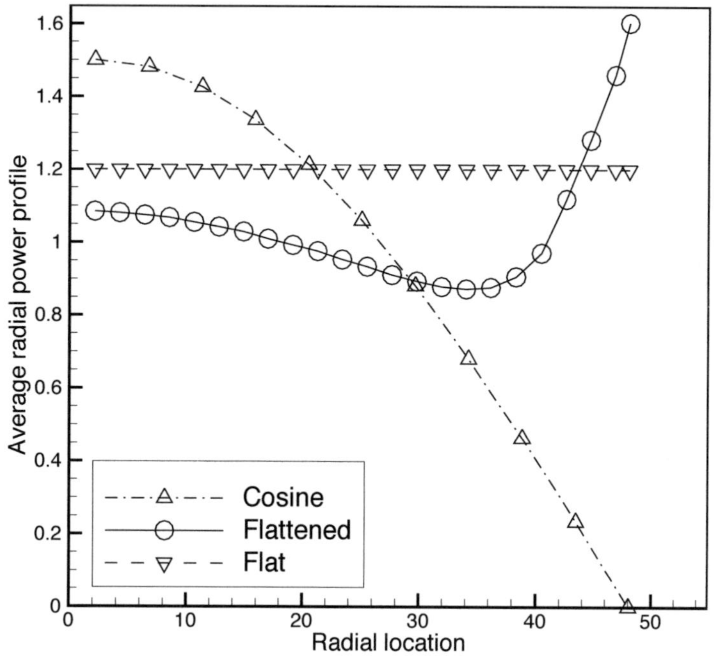

Figure 3. Power profiles employed in the radial direction.

3.2. Thermal properties and kinetics

High-temperature real-gas thermodynamic properties were obtained from [20]. These properties were generated for temperatures up to 20,000 K. The peak gas temperature computed in this study did not exceed 10,000 K, hence is well within the applicable range. A 2-species, 2-reaction chemical kinetics mechanism was used to describe the finite-rate hydrogen dissociation and recombination reactions, as shown in Table 1. The first hydrogen recombination reaction is abridged from a large set of kinetics mechanism developed for kerosene combustion [21], while an irreversible, second reaction [22] is added to describe the hydrogen decomposition. The kinetics of the first hydrogen recombination step have been benchmarked through many kinetic mechanism studies, as described in Ref. [21], while the kinetic rates of the second hydrogen recombination reaction were measured [22]. Note the first reaction is a reversible reaction.

Reaction[a]	*A*	*B*	*E/R*	M	Ref.
$M + H + H = H_2 + M$	5.0E15	0	0	H_2, H	21
$M + H_2 \rightarrow H + H + M$	8.8E14	0	48300	H_2	22

[a]M is third-body collision partner and forward rate constant $K_f = AT^B \exp(-E/RT)$.

Table 1. Hydrogen reaction kinetics mechanism

The importance of finite-rate chemistry was demonstrated in [7], where thermal hydraulics analyses were conducted for the Small Engine with and without finite-rate chemistry. One of the results show that, when the Cosine-Cosine power profile was imposed on the solid-core, the maximum solid temperature calculated with the finite-rate chemistry case was 5369 K, while that for the frozen chemistry case was much higher at 9366 K. This is because the hydrogen decomposition is endothermic. The frozen chemistry freezes molecular hydrogen throughout the thrust chamber and does not allow the hydrogen decomposition to occur, hence an artificially high maximum solid temperature was calculated. It can be seen that without finite-rate chemistry involved, the computed thermal environment and thrust performance are not physical [7].

The solid-core flow element material is assumed to be the (U, Zr)C-graphite composite A, which was tested as flow element material in a legacy reactor [23]. Properties of thermal conductivity, density, and heat capacity as a function of temperature were obtained for (U, Zr)C-graphite composite A from Ref. [23]. Those properties of Beryllium [19] were used for the reflector and slat in the thrust chamber computation. Slat acts as a buffer between the solid-core and the reflector. The thermal properties for the coating layer deposited on the inner wall of flow channels were also obtained from [23].

4. Small engine single flow element modeling

From the material properties point of view, the main cause of mid-section corrosion was speculated as a mismatch in the thermal expansion of flow element and its coating material [3, 4]. The solution is therefore to improve the material properties through materials development [4, 6]. In addition to the materials development, however, we feel further study through the thermal hydraulics analysis of the fundamental reactor design is also important. That is, reduction of the reactor size often started with reducing the diameter of the flow channels, which results in higher aspect ratio, or longer flow channels. According to the Rayleigh line theory, flow with continuous heat addition in a long tube could choke. When that happens, any further heat addition can only serve to reduce the mass flow rate in the tube or, in other words, to jump to another Rayleigh line of lower mass flow [24]. This phenomenon could cause unintended mass flow mal-distribution in the solid-core reactor, resulting in uneven and high local thermal load in the flow element matrix, thereby cracking the coating material. The goal is therefore to compute if choking could occur in one of the flow channels in the Small Engine. To achieve this goal, the worst case was pursued. That is, the Cosine-Cosine power profile which generates peak thermal load in the center of the reactor was assumed. As a result, the flow element located at the center of the solid-core reactor was selected for the computation. In addition, among the three diameters considered for the flow channel of the Small Engine [5], the smallest flow channel diameter was selected, for its highest aspect ratio. These choices were made such that the computed hot hydrogen flow inside one of the flow channels had the best chance to choke.

http://dx.doi.org/10.5772/52569

4.1. Computational grid generation

As mentioned above, the flow element is shaped like a hexagonal prism, containing 19 tubular coolant channels for the propellant hydrogen to acquire heat from the neutronic reactions, as sketched in Fig. 4. For computational efficiency, a 60º pie-section of a single flow element was computed by taking advantage of the symmetrical nature of the prism geometry. A hybrid volume grid was generated by extrusion [25, 26] from a 2-D cross-sectional mesh, as shown in Fig. 5. Structured grid cells were used for the inner and outer layers of the flow channel, and for the outer edges of the prismatic flow element, while unstructured grid cells were used for the rest of the internal web. The structured grid cells at the inner layer of the flow channel was to resolve the turbulent boundary layer, while the general strategy of the hybrid mesh was to minimize numerical diffusion and the number of cells.

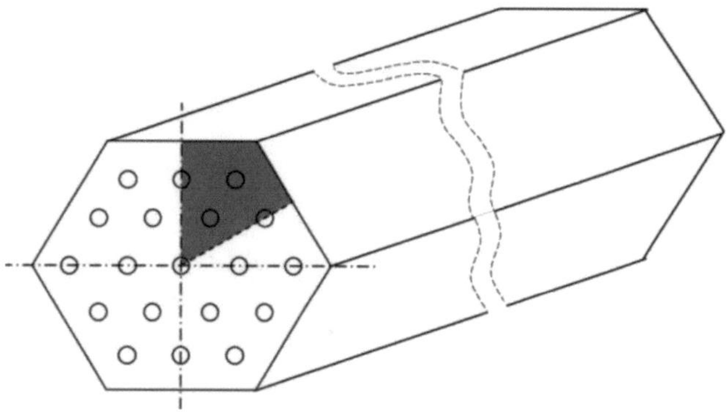

Figure 4. Sketch of the flow element geometry.

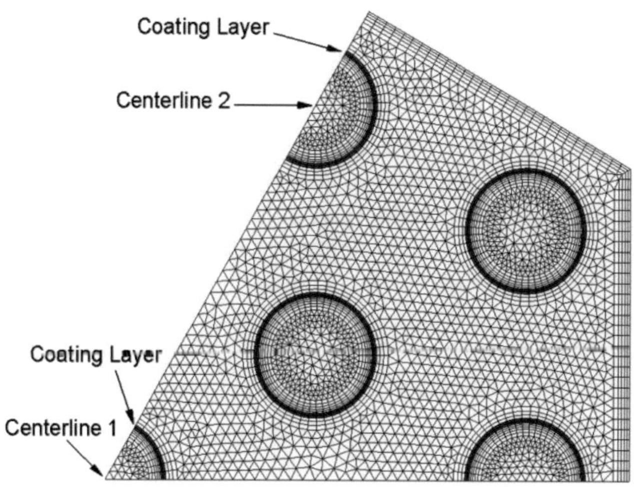

Figure 5. Cross-sectional computational grid layout for the flow channels and the solid web.

In this study, a coating layer, which has different thermal properties from those of the flow element, was also modeled for the flow channel. This was done to imitate the design of those flow elements described in the legacy engine test [23], in which the coating layer was added to protect the carbonaceous compound in the flow element from the alleged chemical attack by the heated, high temperature hydrogen. It is noted that, in the present simulation, the thermal properties of the flow element and the coating layer vary with temperatures. Lastly, in the computational model, an entrance region and an exit region were added to the upstream and downstream of the flow element, respectively, to simulate the environments above and below the solid-core. Total number of cells used was 4.5 million. The heat transfer between the fluid flow, coating layer and the solid fuel was simulated with the conjugate heat transfer model.

4.2. Results and discussion

In this study, various power generation levels were simulated, and it was found that flow choking occurred in some coolant channels as the power level reached about 80% of the maximum power level. Theoretically, once the flow choking occurs, further increase of heat addition would lead to the reduction of mass flow rate in the flow channel, amount to shifting from one Rayleigh line to another as described in Ref. [24]. This could cause mass flow maldistribution in the flow-element matrix, resulting in uneven thermal load in the solid-core. That is, the temperature of the internal web houses the flow channels starved with coolant hydrogen may rise unexpectedly, potentially leading to the cracking of the coating material. Unfortunately, further increase of the power led to unrealistic numerical solutions because of the boundary conditions employed (fixed mass flow rate at the inlet and mass conservation at the exit). This set of boundary conditions was used because only one flow element was simulated, and thus, mass flow reduction at higher power level was precluded. To allow for local mass flow reductions, at least 1/12 of a pie shaped cross-section of the solid-core has to be computed such that fixed mass flow boundary condition need not be used. However, that option was out of the scope of this study. Nevertheless, the current setup is still acceptable since our goal is to determine whether flow choking could occur in the flow channel.

Figure 6 shows the computed temperature distribution in the radial direction (across the solid-fluid interface) at the mid-section of the flow channel. The radial temperature profile reveals that substantial thermal gradients occur at the fluid-coating interface and the coating-solid fuel interface. This kind of high thermal gradient occurring at the coating-solid interfaces could be an issue already without flow choking in the flow channel. With flow choking and the coolant flow rate reduced, the thermal gradient at the coating-solid interface could be even higher since heat is not carried away at the design point, and eventually mid-section corrosion could develop.

Figure 7 shows the computed axial temperature and Mach number profiles along centerlines 1 and 2 (see locations of centerlines 1 and 2 in Fig. 5). The axial temperature profiles almost overlap for centerlines 1 and 2, so are the axial Mach number profiles. The peak Mach number exceeds unity near the end of the flow channel. It can also be seen that the temperatures of propellant hydrogen, rise steadily but drop about 200 K to 2300 K near the end of the fuel port. That is because near the end of fuel port, the flows become choked in both channels. The predicted maximum temperature of both flow channels is about 2700 K. The most significant result from Fig. 7 is that the flow choked near the end of the fuel port, indicating the mass flow rates in these two channels could be reduced, resulting in higher flow element web internal temperature and higher thermal gradient at the coating-solid fuel interfaces and eventually, the possible cracking of the coating layer. Under the operating conditions and assumptions made in this study, the possibility of chocked flow occurring in the flow channels is therefore demonstrated. For future study, three-dimensional numerical simulations of multiple flow elements with a fixed upstream total pressure boundary condition, and a downstream nozzle to remove the fixed exit mass flow boundary condition are necessary to include the inter-element effect. Although the chocked flow occurring in the flow channel is a potential design flaw, the impact of which can readily be avoided by shortening the length of the flow channel to, say, 0.85 m (Fig. 7), without lowering the maximum gas temperature.

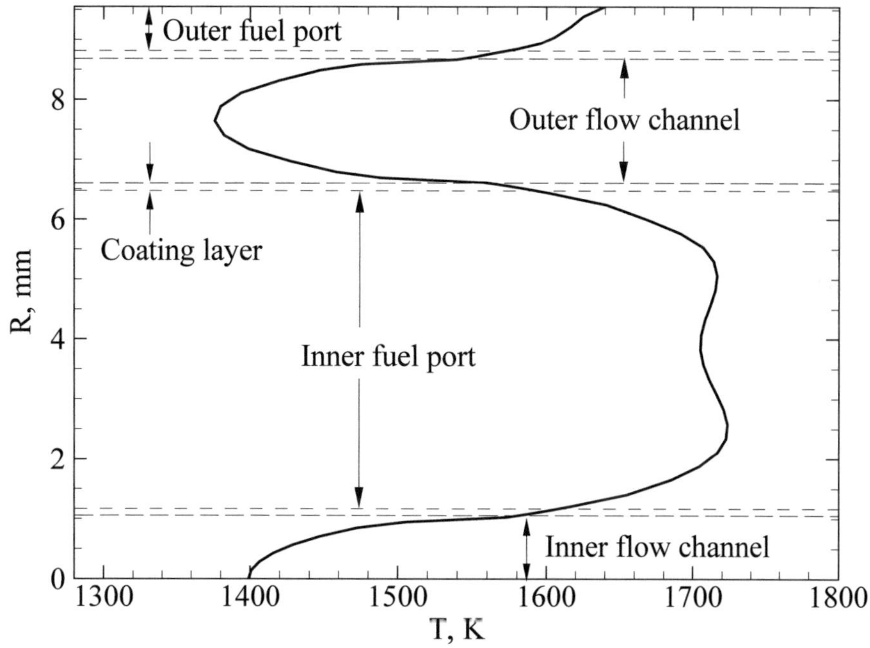

Figure 6. Radial temperature profiles at the mid-section of the flow element.

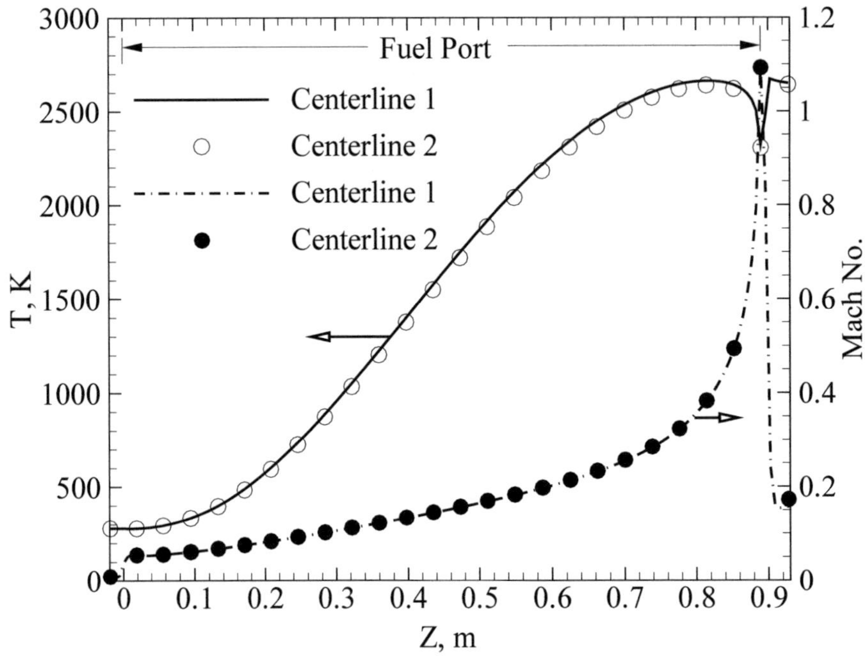

Figure 7. Axial temperature and Mach number profiles along centerlines 1 and 2.

5. Small engine thrust chamber modeling

The rationale of analyzing the entire thrust chamber was to bridge the Small Engine design in the Rover/NERVA era to the current modern thermal hydraulic computational methodology. Specifically, to relate the assigned power profiles to that of the Small Engine by comparing the computed thrust performance with that of the design. Since there are 564 flow elements and 241 support elements or tie-tubes, and each flow element contains 19 tubular flow channels, it is computationally cost-prohibitive to solve detailed thermal hydraulics for each and every flow channel. The 19 flow channels in each flow element were therefore lumped as one porous flow channel for computational efficiency, and the effect of drag and heat transfer were modeled with flow and conjugate heat transfer through porous media, or Eq. (8) through Eq. (12). The medium coolant channel diameter, out of the three coolant channel diameters designed for the Small Engine [5], was selected to derive the porosity of the flow elements.

5.1. Computational grid generation

Hybrid computational grids were generated using a software package GRIDGEN V15.07 [27]. The layout of the flow elements and tie-tubes of the solid-core is such that the whole thrust chamber is symmetrical about a 30 deg. pie-shaped cross-section, hence only the 30

http://dx.doi.org/10.5772/52569

deg. cross-section was computed assuming symmetry. The strategy of using hybrid mesh was again for computational efficiency. Figure 8 shows a 240 deg. view of the computational grid and geometry layout of the Small Engine thrust chamber. The thrust chamber includes the hydrogen inlet duct, pressure vessel, and a nozzle with an expansion ratio of 100. Only tie-tubes in the pressure vessel are shown for clarity. Figure 9 shows a cross-sectional view of the solid-core computational grid, depicting flow elements, tie-tubes, slat, and reflector.

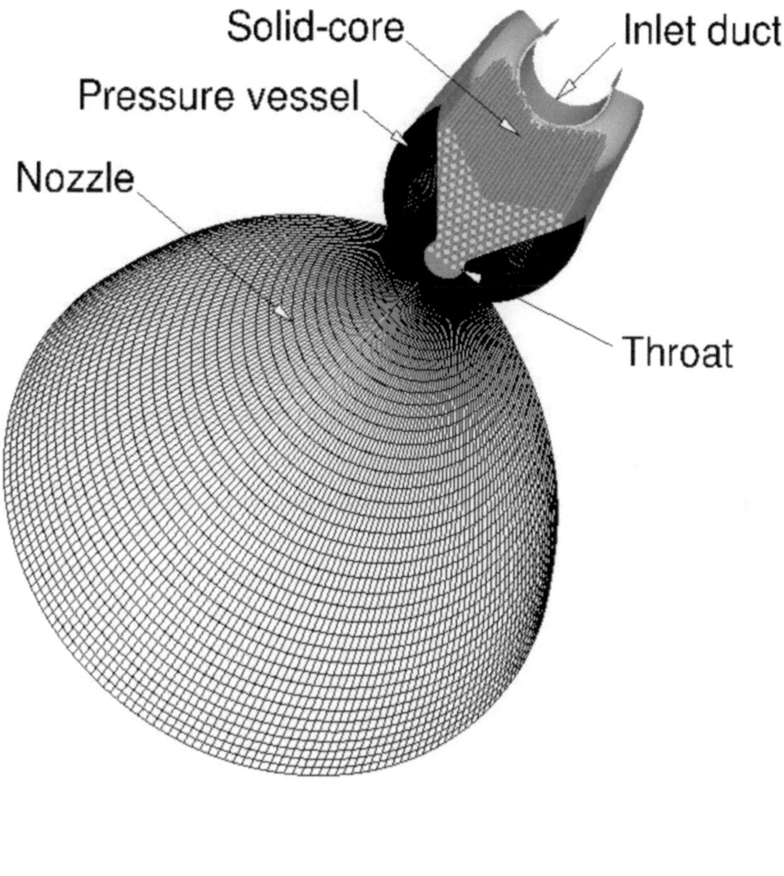

Figure 8. Computational grid and geometry layout of the solid-core Small Engine thrust chamber.

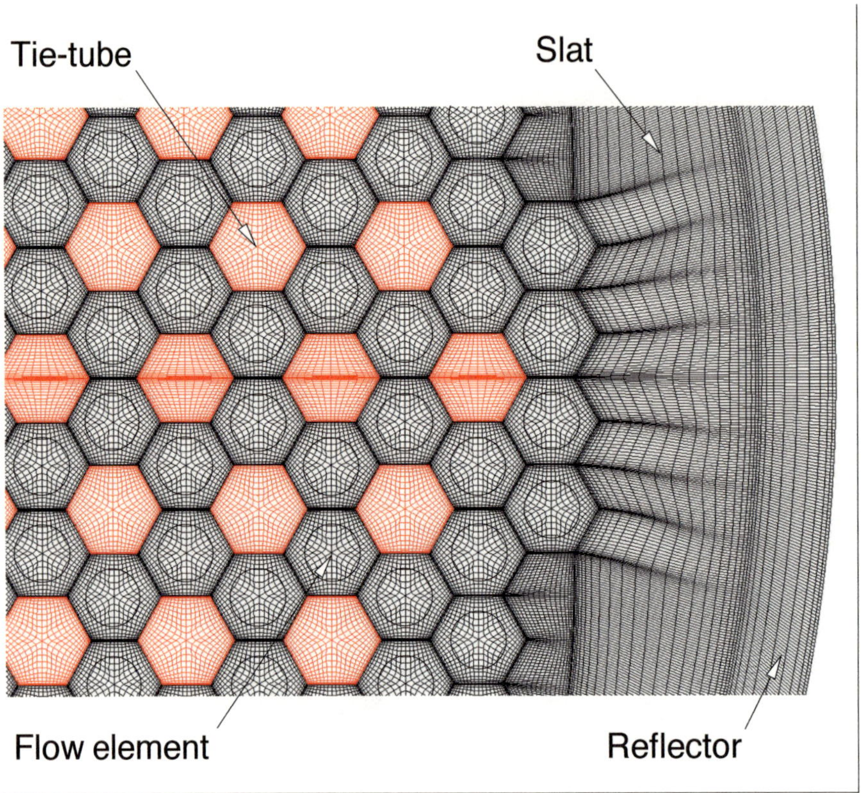

Figure 9. A cross-sectional view of the solid-core. For illustration, circles are imposed onto the flow elements to differentiate them from the tie-tubes.

No-slip boundary condition was applied to all solid walls, while supersonic outflow boundary condition was employed at the nozzle exit. A fixed total pressure and temperature condition was used at the inlet. Inlet hydrogen flow properties were obtained from a system model simulation. Since the minor thermal effects of tie-tubes were included in the system model simulation, an adiabatic wall boundary condition was used for tie-tube walls. The core surrounding components such as the slat and reflector were treated as heat conducting solids to provide accurate boundary conditions for the solid-core boundary. The heat conducted from the core through slat and reflector to the outer thrust chamber walls was dissipated to a far-field temperature assumed to be 310 K. Table 2 shows the run matrix with three combinations of the axial and radial power profiles: the Cosine-Cosine, clipped Cosine-flattened (Cosine), and clipped Cosine-flat profiles. A series of mesh studies using grid sizes of 1,903,278, 1,925,742, 4,681,751, and 7,460,255 points were performed on case 1. It was found that the computed average pressure drops across the solid-core were very similar and the differences among the computed thrust values were less than 2.5%. Based on those findings and the consideration for computational efficiency, the grid size of 1,925,742 points (or 2,408,198 computational cells) was selected for the rest of the computations.

http://dx.doi.org/10.5772/52569

Case	Axial power shape	Radial power shape
1	Cosine	Cosine
2	Clipped Cosine	Flattened
3	Clipped Cosine	Flat

Table 2. Run matrix.

5.2. Results and discussion

5.2.1. Temperature contours and profiles in the thrust chamber

Figure 10 shows the computed thrust chamber temperature contours with three different power profiles: the Cosine-Cosine, clipped Cosine-flattened, and clipped Cosine-flat profiles. In the plotted temperature contours, those in the solid-core represent the solid temperatures in the flow elements; the portions surrounding the solid-core depict the temperatures in the slat and reflector, while the rest are the gas temperature contours. It can be seen that the solid-core temperature contours on the left (case 1) reflect the effect of Cosine-Cosine power distribution, since the peak temperature is located near the middle of the reactor, except it is shifted downstream of the geometrical center in the axial direction. The hydrogen gas temperature contours in the solid core take the same shape as those of the solid temperature, except lower. The heat transfer delay in the axial direction is a result of the energy balance between the cooling from the incoming cold hydrogen in the flow channels and the heating from the nuclear material in the web of the flow elements. It is apparent that the effect of cold hydrogen won the fight between the two counter-acting phenomena, and pushes the peak flow element temperature downstream that shows up as a delay in heat transfer.

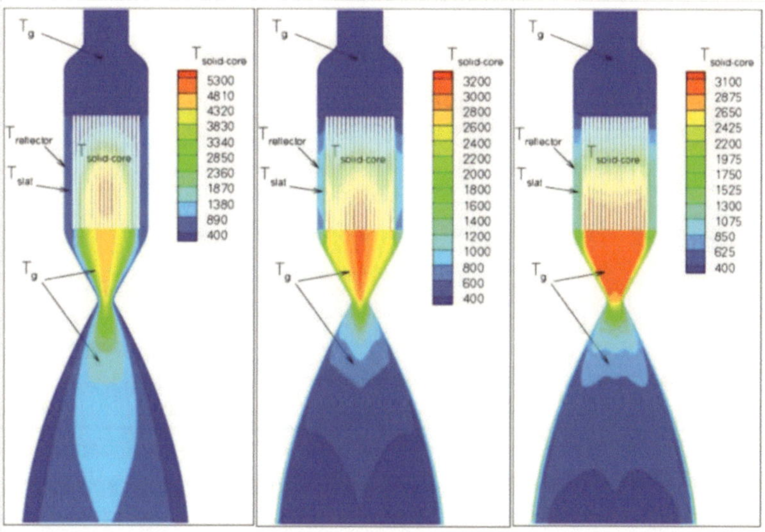

Figure 10. Solid and gas temperature contours on the symmetry plane. From left to right: case 1, case 2, and case 3.

Figure 11 shows the computed solid and gas axial temperature profiles on the symmetry plane, along the centerline of each flow element. Only the temperature profiles for the 1st, 3rd, 5th, 7th, and 9th flow elementfrom the center tie-tube are plotted for clarity. The solid temperatures (T_s shown as dashed lines) lead those of the gas temperatures (T_g, solid lines), showing the effect of the heat transfer lag between the flow element matrix and hydrogen gas. The gas temperatures appear to peak near the core exit, while the solid temperatures appear to peak more upstream for the 1st, 3rd, and 5th flow element, comparing to the peak temperature locations of the hydrogen. The lead decreases at the 7th flow element, and disappears completely at the 9th flow element. These axial temperature profiles reflect the effect of the Cosine-Cosine power profile that concentrates the power in the middle of the core, and drops to zero power at the core boundary. The peak solid temperature for the 9th flow element is the lowest at around 700 K, as expected. The gas temperature at the core-exit is about 4600 K for the 1st flow element, and again about 700 K for the 9th flow element. The 3900 K core-exit gas temperature spread between the 1st and 9th flow elements indicates a very non-uniform temperature distribution at the core-exit, or poor heat transfer efficiency, apparently caused by the power-centric Cosine-Cosine power distribution. Note the temperature gradient of the solid temperature is very steep for the 1st flow element occurring at a location between the core entrance and the peak temperature.

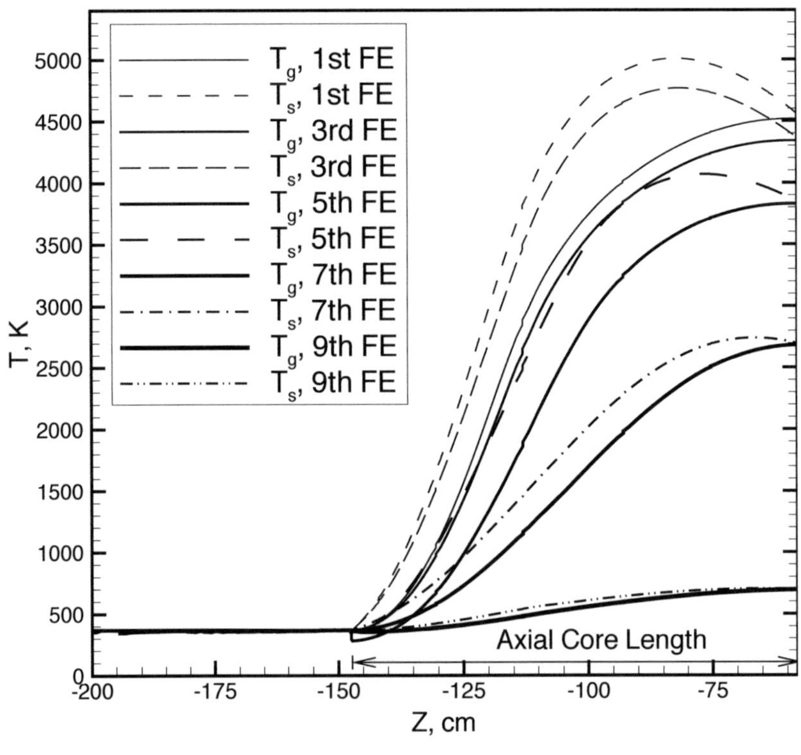

Figure 11. Solid and gas axial temperature profiles for the 1st, 3rd, 5th, 7th, and 9th flow elements on the symmetry plane for case 1.

As discussed above, it is apparent that the Cosine-Cosine power distribution causes not only peak solid temperatures that are too high, but also very non-uniform core-exit gas temperatures, indicating inefficient heat transfer. Fortunately, by measures such as varying the fuel loading, a flattened (Cosine) distribution in the radial direction could be achieved, as indicated in Fig. 3. Note that this profile is not completely flat, since the normalized power is about 1.1 at the center, drops to a minimum of about 0.9 at two-thirds of the radius outward, then rise to a maximum power of 1.6 at the boundary. But comparing it to the Cosine-Cosine power profile, it is relatively flat. In fact, by our estimation, this clipped Cosine–flattened power profile is the most representative of that intended for the Small Engine. The middle picture of Fig. 10 shows the computed solid and gas temperatures, for this clipped Cosine-flattened power profile. It can be seen that the solid and gas temperature contours, reflect the effect of the clipped Cosine-flattened power distribution. Comparing to the temperature contours in the left of Fig. 10, those in the middle are more uniform. The peak solid and gas temperatures in the middle of Fig. 10 at about 3149 and 3081 K., respectively, are much lower than those in the left of Fig. 10, or 5369 and 4596 K., respectively.

Figure 12 shows the computed solid and gas axial temperature profiles on the symmetry plane. Comparing to those of Fig. 11, it can be seen that other than the 9th flow element, the solid temperature generally leads the gas temperature only slightly, since the clipped Cosine-flattened power profile is more uniform than the Cosine-Cosine profile. The peak core-exit gas temperature of the 1st fuel element is about 3037 K., while that of the 9th fuel element is about 2514 K. The range of the gas temperatures at the core-exit is hence only about 523 K, much less than that of 3900 K in case 1, indicating the core-exit gas temperature of case 2 is much more uniform than that of case 1. The flattened radial power profile is therefore a vast improvement over the Cosine radial power distribution.

As mentioned before, it is possible to make the flattened radial power profile in Fig. 3 completely flat, by further fine tuning the fuel loading and by shaping the radial hydrogen mass flow rate profile to match that of the shape of the radial power profile; hence, the proposed theoretical flat radial power profile. The hydrogen mass flow rate profile may be shaped with the installation of various sizes of orifices at the entrance of each flow element. The result of the computed solid and gas temperature contours of such a case, or case 3, using the clipped Cosine-flat power profile, is shown in the right of Fig. 10. It can be seen in terms of the solid-core radial temperature distribution and the uniformity of the gas temperature beneath the solid core, this result is the most uniform. Figure 13 shows the solid and gas axial temperature profiles for the 1st, 3rd, 5th, 7th, and 9th flow elements in the solid-core for case 3. It can be seen that the temperature spread among the 1st, 3rd, 5th and 7th flow elements are much smaller than those of earlier cases, except those of the 9th flow element due to the heat loss to the slat and reflector. From the right picture of Fig. 10 and from Fig. 13, it can be seen this flat radial power profile is a very good power profile in terms of general uniformity of the temperatures for both solid and gas temperatures inside the core and for gas exit temperatures.

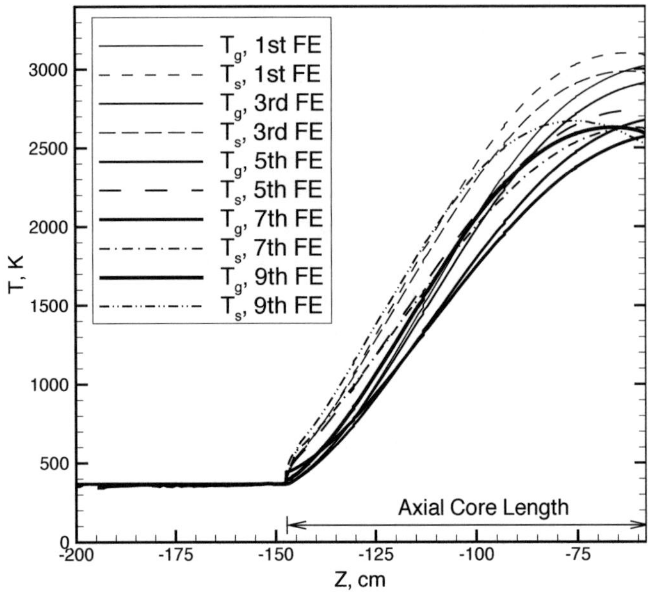

Figure 12. Solid and gas axial temperature profiles for the 1st, 3rd, 5th, 7th, and 9th flow elements on the symmetry plane for case 2.

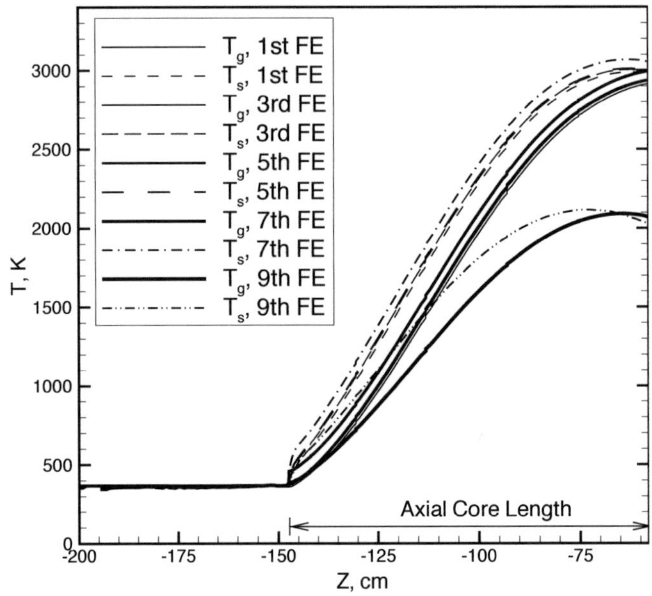

Figure 13. Solid and gas axial temperatures for the 1st, 3rd, 5th, 7th, and 9th flow elements on the symmetric plane for case 3.

http://dx.doi.org/10.5772/52569

5.2.2. *Hydrogen atom mass fraction contours in the thrust chamber*

Figure 14 shows the computed thrust chamber hydrogen mass fraction contours with three aforementioned power profiles: the Cosine-Cosine, clipped Cosine-flattened, and clipped Cosine-flat profiles. The hydrogen atom mass fractions are important because the amount of which is directly related to the eventual temperature contours. As a result, the hydrogen atom mass fraction contours look qualitatively similar to those of the solid-core temperature contours, with the peak conversion pushed to near the end of the core due to the same reason that pushed the peak solid temperature away from the center of the core. For the Cosine-Cosine power profile case, as shown in the left picture of Fig. 14, coolant hydrogen enters the thrust chamber at about 370 K, heats up to about 4800 K in the solid-core, then cools and expands into the diverging nozzle to generate the thrust. Molecular hydrogen decomposes to atomic hydrogen at about 2400 K; hence, most of the hydrogen atoms are formed while in the flow elements. Once the local temperature starts to cool, i.e. during the expansion in the nozzle, hydrogen atoms recombine to become molecular hydrogen. The peak hydrogen atom mass fraction in case 1 is 0.40, or 40% conversion from hydrogen molecule decomposition.

The middle picture in Fig. 14 shows the computed hydrogen atom mass fraction contours for case 2. It surely reflects the effect of the clipped Cosine-flattened power distribution shown in Figs. 2 and 3. Since the peak solid and gas temperatures of the clipped Cosine-flattened power profile are lower than those of the Cosine-Cosine power profile, as displayed in the previous section, the peak hydrogen atom mass fraction also drops from a high of 0.40 in the Cosine-Cosine power profile case to a lowly 0.02, or a 2% conversion in the clipped Cosine-flattened power profile case. Note that the atomic hydrogen contours in case 2 exhibit a more pronounced striation in the pressure vessel beneath the core than those in case 1. This is caused by the higher power profile near the core boundary, resulting in higher local temperatures, thereby higher hydrogen molecule conversion near the core boundary for case 2 than that for case 1.

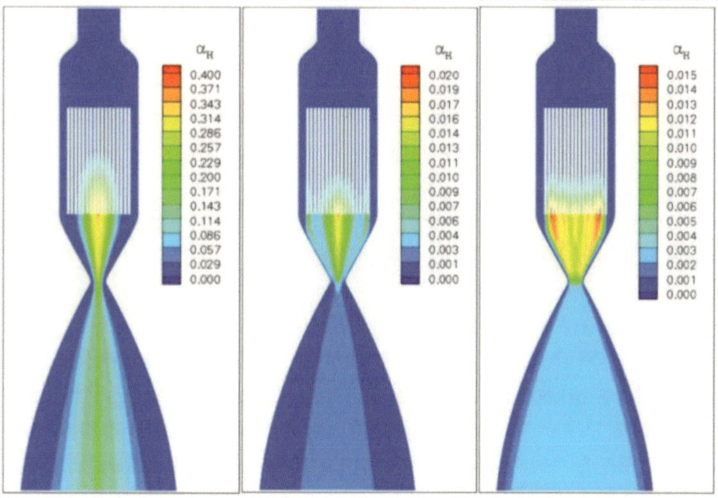

Figure 14. Hydrogen atom mass fraction contours on the symmetry plane. From left to right: case 1, case 2, and case 3.

Shown in the right figure of Fig. 14 is the computed hydrogen atom mass fraction contours for the clipped Cosine-flat power profile case, or case 3. It can be seen that the hydrogen atom mass fraction contours are the most uniform throughout the thrust chamber, among the three cases. The effect of the flat radial power profile is apparently the driver. As a result, the hydrogen atom conversion drops from 2% in case 2 to 1.5% in case 3.

5.2.3. *Summary*

The computed heat transfer and thrust performance design parameters for all three cases are compared with those available of the Small Engine and summarized in Table 3. The computed core pressure drops are shown in the second column. The values for all three cases are close to the design number, although that of the Cosine-Cosine power distribution case is slightly lower. The computed peak solid temperatures are shown in the third column, in which the highest peak solid temperature belongs to case 1, followed by case 2, with the lowest peak solid temperature comes from case 3, as expected.

The fourth column shows the average core-exit gas temperatures. These were obtained by averaging the temperatures at the core-exit for all nine elements on the symmetry plane. It can be seen that the average core-exit gas temperature for the Cosine-Cosine power distribution at 3334 K, is much higher than that of the Small Engine design temperature at 2750 K. On the other hand, the average core-exit gas temperatures for the clipped Cosine-flattened and clipped Cosine-flat power distribution cases are both quite close to that of the Small Engine at 2785 and 2782 K, respectively.

Case	ΔP_{core}, atm	$T_{s,max}$, K	$T_{g,core\ exit}$, K	$\Delta T_{g,core\ exit}$, K	ISP
1	8.9	5369	3334	3900	811
2	9.1	3149	2785	523	868
3	9.1	3066	2782	842	900
Small Engine	9.0	-	2750	-	860~875

Table 3. A summary of the heat transfer and performance parameters.

The computed temperature spread, or the difference of the core-exit gas temperature between the first and the 9th flow elements, is shown in the fifth column. Lower temperature spread represents more uniform temperature distribution amongst the flow elements, or better heat transfer efficiency. It can be seen that the clipped Cosine-flattened power profile produces best heat transfer efficiency, the clipped Cosine-flat power profile ranks the second, while the Cosine-Cosine power profile has the worst heat transfer efficiency. Note that if without the outlier of the temperature profile at the 9th fuel element for case 3, its temperature spread from the rest is more uniform than that of case 2, as indicated in Figs. 12 and 13.

http://dx.doi.org/10.5772/52569

From the last column comes with the comparison of the specific impulses. The best thrust performance belongs to case 3 that is much higher than the design value, followed by case2 which is within the design range, while case 1 has the lowest ISP. Yet the maximum atomic hydrogen conversions are 40%, 2% and 1.5%, for cases 1, 2, and 3, respectively. These results demonstrate that the notion of higher temperature, more atomic hydrogen, therefore higher thrust is not valid. Rather, it is the most uniform radial power profile that produces the highest thrust.

In summary, the computed core pressure drop, core-exit gas temperature, and specific impulse, from the clipped Cosine-flattened power distribution agree well with those of the Small Engine. The result indicates the fluid, thermal, and hydrogen environments computed with the present computational thermal hydraulics methodology closely emulate what was intended for the original Small Engine design. In addition, examining the results of our proposed clipped Cosine-flat power distribution, not only the computed core-exit gas temperature and core pressure drop agree equally well with those of the Small Engine, the computed specific impulse is 25 to 40 seconds higher than those of the Small Engine, demonstrating the present computational methodology can be used to guide the future design.

6. Conclusion

Thermal hydraulics computational design analyses were performed to investigate the mid-section corrosion issue occurred during the legacy engine tests, and to predict the heat transfer efficiency and thrust performance for a virtual solid-core, nuclear thermal engine thrust chamber – the Small Engine. Multiphysics invoked include the turbulent flow and heat transfer, finite-rate chemistry, power generation, and conjugate heat transfer for solids and porous media. The results of a detailed single flow element modeling show that under the assumption of the worst design condition, the hot hydrogen flow could choke near the end of the narrow, long flow channels, which could lead to hydrogen flow reduction and local hot spots in the solid-core; hence, originating the mid-section corrosion. In addition, a unified Small Engine thrust chamber thermal flow field constituting those of the inlet plenum, the pressure vessel, and the nozzle, was analyzed with three power profiles. It was found that the computed core pressure drop, core-exit gas temperature, and specific impulse for the clipped Cosine-flattened power profile closely agreed with those of the Small Engine design values. It was also found that with our proposed clipped Cosine-flat power profile, not only the computed core-ext gas temperature and core pressure drop closely agreed with those of the Small Engine, the corresponding specific impulse was much higher than those of the original Small Engine numbers. Design lesson learned from this effort is that high hydrogen molecule conversion to hydrogen atoms neither improves the heat transfer efficiency, nor increases the thrust performance. Rather, it is the more uniform radial power profile that produces lower peak solid temperature, higher heat transfer efficiency, and higher thrust performance.

Nomenclature

A=pre-exponential rate constant

B=temperature power dependence

C_1, C_2, C_3, C_μ=turbulence modeling constants, 1.15, 1.9, 0.25, and 0.09

C_p=heat capacity

D=diffusivity

d=flow channel diameter

E=activation energy

f_L, f_q=empirical multipliers

H=total enthalpy

K=thermal conductivity

k=turbulent kinetic energy

L=drag loss due to porous media

P=pressure

p_r=Prandtl number

Q=volumetric heat source

R=radial distance

R=gas constant

Re=Reynolds number

T=temperature

t=time, s

U=flow speed

u_i=mean velocities in three directions

x=Cartesian coordinates

Z=axial distance

α=species mass fraction

β=porosity or void of fraction

ε=turbulent kinetic energy dissipation rate

μ=viscosity

μ_t=turbulent eddy viscosity (=$\rho C_\mu k^2/\varepsilon$)

Π=turbulent kinetic energy production

ρ=density

σ=turbulence modeling constants

τ=shear stress

ω=chemical species production rate

Subscripts and superscripts

cl=centerline

p=nuclear power source

r=radiation

s=surface, solid, or porous media

t=turbulent flow

Acknowledgments

This study was partially supported by a Nuclear Systems Office task entitled "Multiphysics Thrust Chamber Modeling" with funding from the Prometheus Power and Propulsion Office, NASA Headquarter. Ron Porter was the Marshall Space Flight Center Nuclear Systems Office Manager. Wayne Bordelon was the Nuclear Thermal Propulsion manager. Michael Houts was the Nuclear Research Manager. Steve Simpson and Karl Nelson provided the clipped Cosine-flattened power profile, nozzle geometry and system modeling results. Bill Emrich suggested the Cosine-Cosine power profile. Solid material thermal properties provided by Panda Binayak, Robert Hickman, and Bill Emrich are also acknowledged.

Author details

Ten-See Wang[1], Francisco Canabal[1], Yen-Sen Chen[2], Gary Cheng[3] and Yasushi Ito[3]

1 NASA Marshall Space Flight Center, Huntsville, Alabama, USA

2 Engineering Sciences Incorporated, Huntsville, Alabama, USA

3 University of Alabama at Birmingham, Birmingham, Alabama, USA

References

[1] Bordelon, W.J., Ballard, R.O., and Gerrish, Jr., H.P., "A Programmatic and Engineering Approach to the Development of a Nuclear Thermal Rocket for Space Exploration," AIAA Paper 2006-5082, July 2006.

[2] Howe, S.D., "Identification of Archived Design Information for Small Class Nuclear Rockets," AIAA Paper 2005-3762, 41st AIAA/ASME/SAE/ASEE Joint Propulsion Conference, Tucson, Arizona, 2005..

[3] Emrich, W.J., Jr., "Non-Nuclear NTR Environment Simulator," Space Technology and Applications International Forum (STAIF-2006), Albuquerque, NM, Feb. 12-16, 2006, American Institute of Physics Proceedings, edited by El-Genk, M.S., Melville, N.Y., Vol. 813, 2006, pp. 531-536.

[4] Wang, T.-S., Luong, V., Foote, J., Litchford, R., and Chen, Y.-S., "Analysis of a Cylindrical Specimen Heated by an Impinging Hot Hydrogen Jet," Journal of Thermophysics and Heat Transfer, Vol. 24, No. 2, April-June, 2010, pp. 381~387. doi: 10.2514/1.47737.

[5] Durham, F.P., "Nuclear Engine Definition Study", Preliminary Report, Vol. 1 – Engine Description, LA-5044-MS, Los Alamos Scientific Laboratory, Los Alamos, New Mexico, September 1972.

[6] Wang, T.-S., Foote, J., and Litchford, R., "Multiphysics Thermal-Fluid Design Analysis of a Non-Nuclear Tester for Hot-Hydrogen Material Development," Space Technology and Applications International Forum (STAIF-2006), Albuquerque, NM, Fe. 12-16, 2006, American Institute of Physics Proceedings, edited by El-Genk, M.S., Melville, N.Y., Vol. 813, 2006, pp. 537-544.

[7] Wang, T.-S., Canabal, F., Chen, Y.-S., Cheng, Gary, "Multiphysics Computational Analysis of a Solid-Core Nuclear Thermal Engine Thrust Chamber," Journal of Propulsion and Power, Vol. 26, No. 3, May-June, 2010, pp. 407~414. doi: 10.2514/1.47759.

[8] Cheng, G., Ito, Y., Ross, D., Chen, Y.-S., and Wang, T.-S., "Numerical Simulations of Single Flow Element in a Nuclear Thermal Thrust Chamber," AIAA Paper 2007-4143, 39th AIAA Thermophysics Conference, Miami, FL, 2007.

[9] Shang, H.M., and Chen, Y.-S., "Unstructured Adaptive Grid method for Reacting Flow Computation," AIAA Paper 1997-3183, Seattle, WA (1997).

[10] Wang, T.-S., Droege, A., D'Agostino, M., Lee, Y.-C., and Williams, R., "Asymmetric Base-Bleed Effect on Aerospike Plume-Induced Base-Heating Environment," Journal of Propulsion and Power, Vol. 20, No. 3, 2004, pp. 385-393.

[11] Wang, T.-S., Chen, Y.-S., Liu, J., Myrabo, L.N., and Mead, F.B. Jr., "Advanced Performance Modeling of Experimental Laser Lightcraft," Journal of Propulsion and Power, Vol. 18, No. 6, 2002, pp. 1129-1138.

[12] Wang, T.-S., "Multidimensional Unstructured-Grid Liquid Rocket Engine Nozzle Performance and Heat Transfer Analysis," Journal of Propulsion and Power, Vol. 22, No. 1, 2006, pp. 78-84.

[13] Wang, T.-S., "Transient Three-Dimensional Startup Side Load Analysis of a Regeneratively Cooled Nozzle," Shock Waves – An International Journal on Shock Waves, Detonations and Explosions. Vol. 19, Issue 3, 2009, pp. 251~264. DOI: 10.1007/s00193-009-0201-2.

[14] Chen, Y.-S., and Kim, S. W., "Computation of Turbulent Flows Using an Extended k-Turbulence Closure Model," NASA CR-179204, 1987.

[15] Bird, R.B., Stewart, W.E., Lightfoot, E.N., "Transport Phenomena," 1960.

[16] Gaski, J., "The Systems Improved Numerical Differencing Analyzer (SINDA) Code – a User's Manual," Aerospace Corp., El Segundo, CA, Feb. 1986.

[17] Wang, T.-S. and Schutzenhofer, L. A., "Numerical Analysis of a Nuclear Fuel Element for Nuclear Thermal Propulsion," AIAA Paper 91-3576, September, 1991.

[18] Cheng, G.-C., Chen, Y.-S., and Wang, T.-S., "Flow Distribution Around the SSME Main Injector Assembly Using Porosity Formulation," AIAA Paper 95-0350, January, 1995.

[19] Glasstone, S., and Edlund, M.C., "The Elements of Nuclear Reactor Theory," D. Van Nostrand Company, Inc., Tornoto, Canada, 1958.

[20] McBride, B.J., Zehe, M.J., and Gordon, S., "NASA Glenn Coefficients for Calculating Thermodynamic Properties of Individual Species," NASA TP-2002-211556, Glenn Research Center, Cleveland, Ohio. September, 2002.

[21] Wang, T.-S., "Thermophysics Characterization of Kerosene Combustion," Journal of Thermophysics and Heat Transfer, Vol. 15, No. 2, 2001, pp. 140-147.

[22] Baulch, D.L., Drysdale, D.D., Horne, D.G., and LLoyd, A.C., "Evaluated Kinetic Data for High temperature Reactions", Vol. 1, The Chemical Rubber Company, Cleveland, Ohio, 1972.

[23] Lyon, L. L., "Performance of (U, Zr)C-Graphite (Composite) and of (U, Zr)C (Carbide) Fuel Elements in the Nuclear Furnace 1 Test Reactor," LA-5398-MS, Los Alamos Scientific Laboratory, Los Alamos, New Mexico, 1973.

[24] James, E.A., J., "Gas Dynamics", Allyn and Bacon, Inc., Boston, Mass., 1969.

[25] Ito, Y., and Nakahashi, K., "Direct Surface Triangulation Using Stereolithography Data," AIAA Journal, Vol. 40, No. 3, 2002, pp. 490-496. DOI: 10.2514/2.1672.

[26] Ito, Y., and Nakahashi, K., "Surface Triangulation for Polygonal Models Based on CAD Data," International Journal for Numerical Methods in Fluids, Vol. 39, Issue 1, 2002, pp. 75-96. DOI: 10.1002/fld.281.

[27] Steinbrenner, J.P., Chawner, J.R., and Fouts, C., "Multiple Block Grid Generation in the interactive Environment," AIAA Paper 90-1602, June 1990.

Chapter 5

CFD Simulation of Flows in Stirred Tank Reactors Through Prediction of Momentum Source

Weidong Huang and Kun Li

Additional information is available at the end of the chapter

http://dx.doi.org/10.5772/51754

1. Introduction

The mixing and agitation of fluid in a stirred tank have raised continuous attention. Starting with Harvey and Greaves, Computational Fluid Dynamics (CFD) has been applied as a powerful tool for investigating the detailed information on the flow in the tank [1-2]. In their work, the impeller boundary condition (IBC) approach has been proposed for the impeller modeling, in which the flow characteristics near the impeller are experimentally measured, and are specified as the boundary conditions for the whole flow field computation [1-4]. Because it depends on the experimental data, IBC can hardly predict the flow in the stirred tank and its applicability is inherently limited. To overcome this drawback, the multiple rotating reference frames approach (MRF) has been developed, in which the vessel is divided into two parts: the inner zone using a rotating frame and the outer zone associated with a stationary frame, for a steady state simulation. Although it can predict the flow field, the computation result is slightly lack of accuracy and needs a longer time for convergence [5]. Sliding mesh approach (SM) is another available approach, in which the inner grid is assumed to rotate with the impeller speed, and the full transient simulations are carried out [6-7]. SM approach gives an improved result, but it suffers from the large computational expense [8]. Moving-deforming grid technique was proposed by Perng and Murthy [9], in which the grid throughout the vessel moves with the impeller and deforms. This approach requires a rigorous grid quality and the computational expenses are even higher than SM [10-11].

Momentum source term approach adds momentum source in the computational cells to represent the impeller propelling and the real blades are ignored. In the approach, the generations of the vessel configuration and grids are simpler, and the computational time is shorter and the computational accuracy is higher. However, the determination of the momentum source depends on experimental data or empirical coefficients presently [12-13]. The approach proposed by Pericleous and Patel is based on the airfoil aerodynamics and originally aimed at the two-dimensional flow in the stirred vessels. Xu, McGrath [14] and Patwardhan [15] applied this approach to simulate the three-dimensional flow pattern in a tank with the pitched blade turbines. Revstedt et al. [16] modified the approach for the three-dimensional simulation in a Rushton turbine stirred vessel. In their approach, the determination of the momentum source depends on the specified power number. Dhainaut et al. [13] reported a kind of momentum source term approach in which the fluid velocity is linearly proportional to the radius with an empirical coefficient. In our previous study, we proposed to calculate the momentum source term according to the ideal propeller equation [17], which is related to the rotation speed and radius of the blade [18-20], however, the prediction accuracy is just a little better than MRF method.

Besides, other methods, such as inner-outer approach, snapshot approach and adaptive force field technique, have been developed, but are less applied[1, 21].

In this study, an equation is proposed to calculate the momentum source term after considering both impeller propelling force and the radial friction effect between the blades and fluid. The flow field in the Rushton turbine stirred vessel was simulated with the CFD model. The available experimental data near the impeller tip and in the bulk region were applied to validate this approach. Moreover, the comparisons of the computational accuracy and time with MRF and SM were carried out.

2. Model and methods

2.1. Stirred tank and grid generation

Fig.1 shows the geometry of the investigated standard six-bladed Rushton turbine stirred tank with four equally spaced baffles. The diameter of cylindrical vessel $T = 0.30$ m, the diameter of rotating shaft $d = 0.012$ m, the hub diameter $b = 0.020$ mm, and the detailed size proportions are listed in Table 1. The working fluid is water with density $\rho = 998.2$ kg m^{-3} and viscosity $\mu = 1.003\times10^{-3}$ Pa s. The rotational speed of the impeller $N = 250$ rpm, leading to a tip speed $u_{tip} = 1.31$ m s^{-1} and Reynolds number Re = 41,467 ($\mathrm{Re} = \rho ND^2/\mu$). Eight axial locations in the bulk region are also shown in Fig.1. They are the same as the experimental study of Murthy and Joshi [22],

It is well known that sufficiently fine grids and lower-dissipation discretization schemes can significantly reduce the numerical errors [23]. And the grid resolution on the blades has an important influence on capturing the details of flow near the impeller [23-24]. Grid independence study has been carried out for momentum source term approach with the standard k-ε model with different grid resolutions. In this paper, the results of the grid resolution of 1,429,798 hexahedral cells ($r\times\theta\times z \approx 88\times120\times131$) with the impeller blade covered with 96000 cells ($r\times\theta\times z = 32\times120\times25$) have been reported for the simulations of standard k-ε model and Reynolds stress model (RSM). In the region encompassing the impeller discharge stream, contained within 1.5 blade heights above and below the impeller and extending horizontally across the tank, the grid was refined to resolve the large flow gradients (Fig.2). For it is unnecessary to construct the impeller in momentum source term approach, the tank geometry and mesh generations are simpler than MRF and SM.

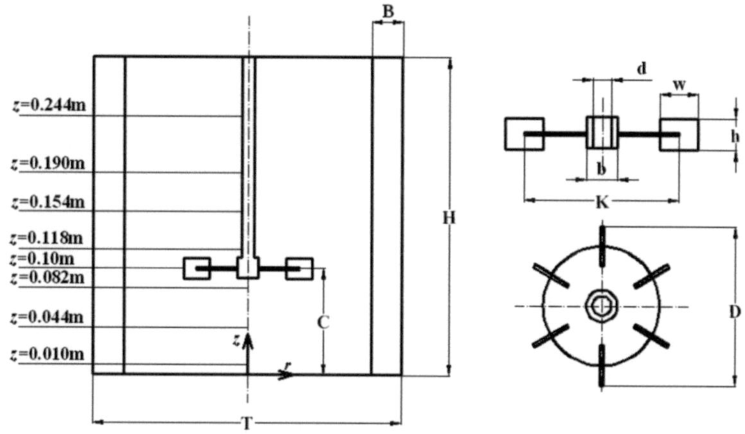

Figure 1. Geometry of the stirred vessel.

H/T	C/T	B/T	D/T	K/D	w/D	h/D
1	1/3	1/10	1/3	3/4	1/4	1/5

Table 1. Dimension scaling of the stirred vessel configuration.

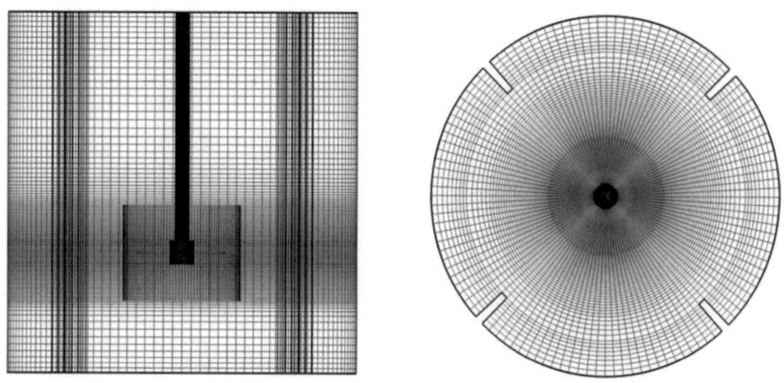

Figure 2. Computational mesh for the momentum source term approach simulation.

In order to investigate the computational speed of MRF and SM, they are also applied to simulate the flow field of stirred tank. The total number of computational cells for MRF is 1,652,532 ($r\times\theta\times z \approx 63\times204\times126$) with the impeller blade covered with 88128 cells ($r\times\theta\times z$ =

24×204×18), and for SM is 606,876 ($r\times\theta\times z \approx 60\times106\times90$) with the impeller blade covered with 33920 cells ($r\times\theta\times z = 20\times106\times16$).

2.2. Momentum source model and control equations

Following the assumptions of Euler equation for turbomachinery [25], the momentum source term from the driving of blade is determined as following. For a small area of blade dS, assuming that a force exerted by the impeller on the fluid is perpendicular to the blade surface [14], the propelling force from the impeller dF is considered to be equal to the product of the mass flow rate across the interaction section dQ and the fluid velocity variation along the normal direction of the area element dS:

$$\mathrm{d}F = \mathrm{d}Q \cdot \Delta \vec{u} \tag{1}$$

Since the rigid body rotational motion of the blade is considered, it is a reasonable assumption that the fluid velocity after being acted on is the same as the velocity of the impeller blade, thus Δu can be obtained. For the present study on the vertical blade, Eq.(1) can be written as:

$$\mathrm{d}F = \rho \cdot \mathrm{d}S \cdot u \cdot (u - v_\theta) \tag{2}$$

where u is linear velocity of the area element on the blade surface dS, v_θ is fluid tangential velocity rotating with the impeller before the fluid was propelled by impeller, dS is cross-section area of the interface between fluid and the impeller blade. A similar equation has been applied to describe the propelling effect of the ship's impeller [26] which we applied it to calculate the momentum source term [18-19].

Furthermore, the fluid is continuously impelled out from the impeller region, so there is a relative motion of the fluid along the radial direction of the blade. In order to consider the friction effect on the fluid movement, the friction resistance equation about the finite flat plate based on the boundary layer theory is introduced to calculate the friction force approximately [27]:

$$\mathrm{d}f = \frac{1}{2}\rho v_r^2 \cdot C_f \cdot \mathrm{d}S \tag{3}$$

where f is friction force in the computational cells; v_r is radial velocity of the fluid; C_f is local resistance coefficient related to Re_x, which is calculated approximately by:

$$\begin{aligned} &C_f = 0.664 \cdot \mathrm{Re}_x^{-0.5} \\ &\mathrm{Re}_x <= 5\times10^5 \end{aligned} \tag{4}$$

$$C_f = 0.0577 \cdot \mathrm{Re}_x^{-0.2}$$
$$5 \times 10^5 < \mathrm{Re}_x < 1 \times 10^7 \tag{5}$$

where $\mathrm{Re}_x = \rho v_r x / \mu$, x is distance between cell center and the center of rotation axis. In this paper, we mainly consider the radial friction resistance, ignoring the axial effect for simplicity. Adding the momentum sources of the both direction into the computational cells, the real blade is replaced.

In the previous study, the difference between the ensemble-averaged flow field calculated with the steady-state and the time-dependent approaches was found to be negligible [28-30]. Here, the continuity and momentum equations of motion for three-dimensional incompressible flow, as well as the standard k–ε turbulence equation [31] or Reynolds stress transport equations [32], were solved to calculate single phase flow of the stirred vessel.

The free surface was treated as a flat and rigid lid, so a slip wall was given to the surface. The disc, hub and shaft of the Rushton turbine are specified as moving walls. A standard wall function was given to the other solid walls, including the bottom surface, the sidewall and baffles. Second order upwind discretization scheme was adopted for pressure interpolation and the convection term of momentum, turbulent kinetic energy and energy dissipation rate equations, and the discretized equations were solved iteratively by using the SIMPLE algorithm for pressure-velocity coupling.

2.3. Power number prediction

The power number N_P is an important parameter of the stirred tank, which is generally applied to validate the CFD predictions [33-34]. One method for calculating the power number is based on the energy balance, in which the power number of the impeller is calculated from the integration of the turbulent energy dissipation rate (ε) predicted from the CFD model [35]:

$$N_P = \frac{\int_0^H \int_0^{2\pi} \int_0^{T/2} \rho \varepsilon r dr d\theta dz}{\rho N^3 D^5} \tag{6}$$

In MRF and SM approaches, the power number is usually calculated from the predicted torque [36]:

$$N_p = \frac{2\pi N m M}{\rho N^3 D^5} \tag{7}$$

in which m is the number of the blades, M is the torque of each blade.

In the present study, since the force from blade results in the fluid movement, the impeller power can be calculated from the integration of the momentum source in the impeller region [14], which is called integral power based approach, so the power number is calculated in the following manner in our study:

$$N_p = \frac{\int_{C-h/2}^{C+h/2}\int_0^{2\pi}\int_{D/2-w}^{D/2}(F_\theta \cdot 2\pi Nr) r \mathrm{d}r \mathrm{d}\theta \mathrm{d}z}{\rho N^3 D^5} \tag{8}$$

in which F_θ is tangential momentum source term.

3. Results and discussion

3.1. Numerical validation of the flow field near the impeller tip

Fig.3 shows the profiles of the predicted and experimental flow data in the impeller region. Fig.3a gives the comparison of the radial velocity. Escudie and Line [37] summarized the previous experimental works and found due to the differences of the experimental technique and the stirred-vessel configuration, there existed some inconsistencies among the reported results, whereas the maximum of radial velocity was in the range of 0.7-0.87u_{tip}. Their experiment study gave a maximal radial velocity of 0.80u_{tip}, while Wu and Patterson [38] 0.75u_{tip}. In present work, momentum source term approach with standard k-ε and RSM predicts a maximum radial velocity of 0.79u_{tip} and 0.75u_{tip}, respectively, which means that momentum source term approach predicts the maximal radial velocity rather well. From fig. 3a, the distribution of radial velocity predictions based on momentum source term approach agrees rather well with the experimental data, which outperforms MRF predictions.

With regard to tangential velocity, the maxima from Wu and Patterson [38], and Escudie and Line [37] are 0.66u_{tip} and 0.72u_{tip} respectively. For momentum source term approach with standard k-ε and RSM, both gave a maximum of 0.63u_{tip}. From Fig. 3b, it can be found that the calculated results of momentum source term approach are in good agreement with the two sets of experimental data, better than MRF predictions.

Fig.3c shows the comparison of axial velocity. The measured results from Wu and Patterson [38], Escudie and Line [37] are almost the same. The momentum source term approach results match the measured data well in most regions, but predicting the change from the maximum to minimum with several deviations. Compared to standard k-ε model, RSM predictions of momentum source term approach make some improvements, while MRF predictions are not provided.

In fig.3d, it can be observed that there are two maxima of different magnitudes in the profiles of turbulent kinetic energy, thus the curve is not symmetrical. RSM simulations of momentum source term approach are in accordance with those of standard k-ε model. Momentum source term approach and MRF both successfully predict the variations between two maxima, but momentum source term approach exhibits a better prediction of the turbulent kinetic energy than MRF.

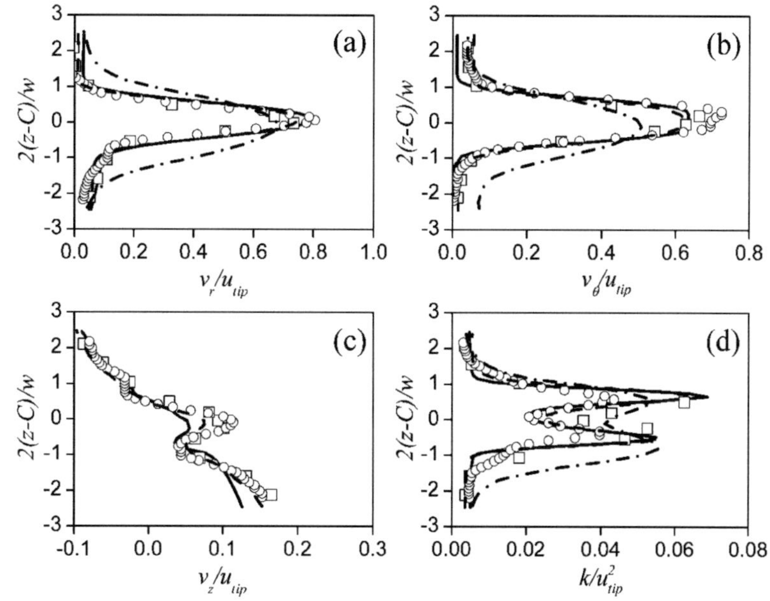

Figure 3. Comparison of the results predicted by different impeller approaches and experimental data around impeller tip (a) radial velocity, (b) tangential velocity, (c) axial velocity and (d) turbulence kinetic energy: ○ experimental data (Escudie and Line, 2003), ▪ experimental data (Wu and Patterson, 1989), standard *k*-ε momentum source term approach predictions, RSM momentum source term approach predictions, MRF predictions (Deglon and Meyer, 2006)

3.2. Numerical validation of the flow field in the bulk region

Fig.4 shows the radial profiles of the mean axial velocity at various axial levels. Here, $z<0.10$ m is in the lower parts of the tank, while $z>0.10$ m is in the upper parts. It can be noted that curve changes in two parts are just opposite, indicating that axial velocity directions are opposite. The result shows that the predictions of momentum source term approach with standard *k*-ε model and RSM are both in favorable agreement with the experimental data, similar to the predicted results of SM.

Fig.5 depicts the comparison of predicted and experimental radial velocity component. The high speed impeller discharge streams radially. Radial velocity initially increases and then decreases, attaining the maximum at r/R = 0.6-0.8 near blade. It can be seen that compared to the experimental data, both the standard *k*-ε model predictions of momentum source term approach and SM exhibit some disparity, particularly at the levels in the upper part. At the top axial levels (z = 0.154 m and z = 0.244 m), there are the largest differences between the predictions of momentum source term approach and SM, while they have similar accuracy in other positions. However, the results of momentum source term approach and SM predicted with RSM are consistent with the experimental data.

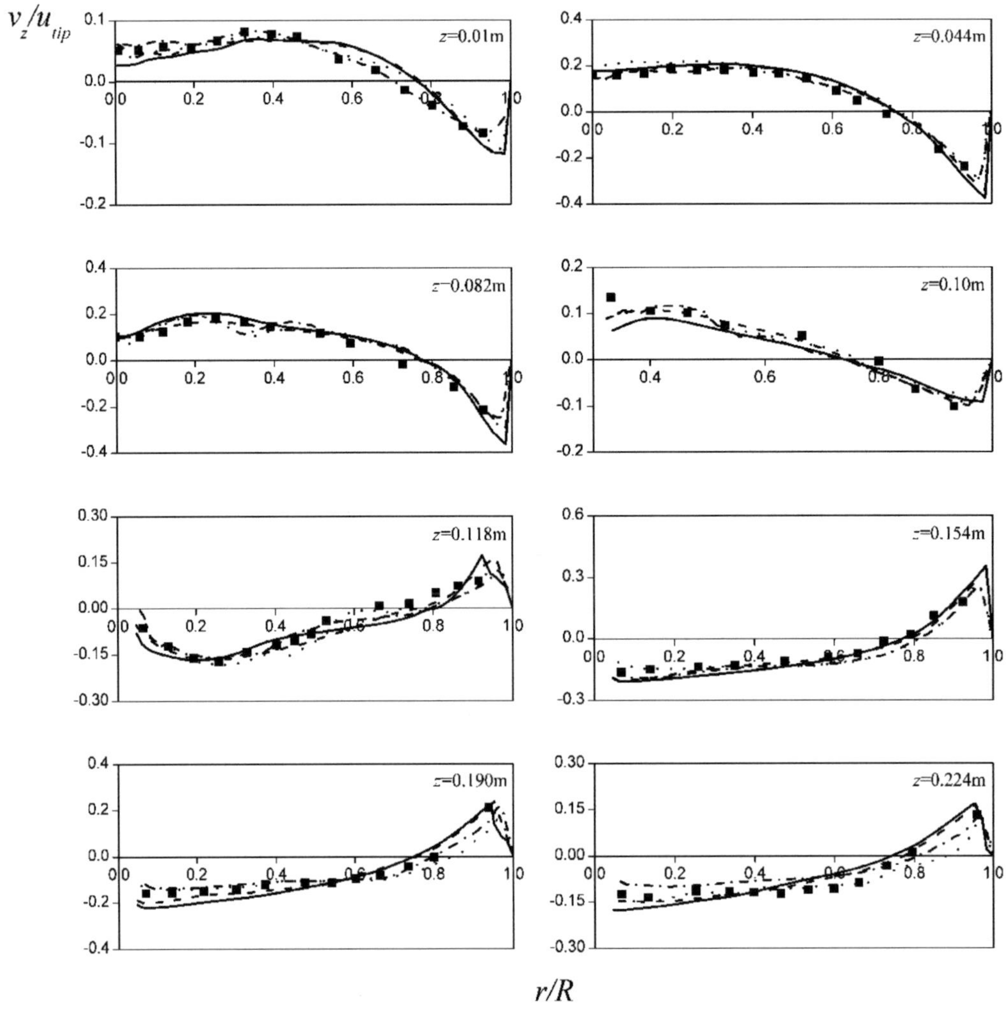

Figure 4. Comparison of the results predicted by different impeller approaches and experimental data of the dimensionless mean axial velocity in the bulk region: experimental data [39] (Murthy and Joshi, 2008) standard k-ε momentum source term approach predictions, RSM momentum source term approach predictions, standard k-ε SM predictions (Murthy and Joshi, 2008), RSM SM predictions (Murthy and Joshi, 2008).

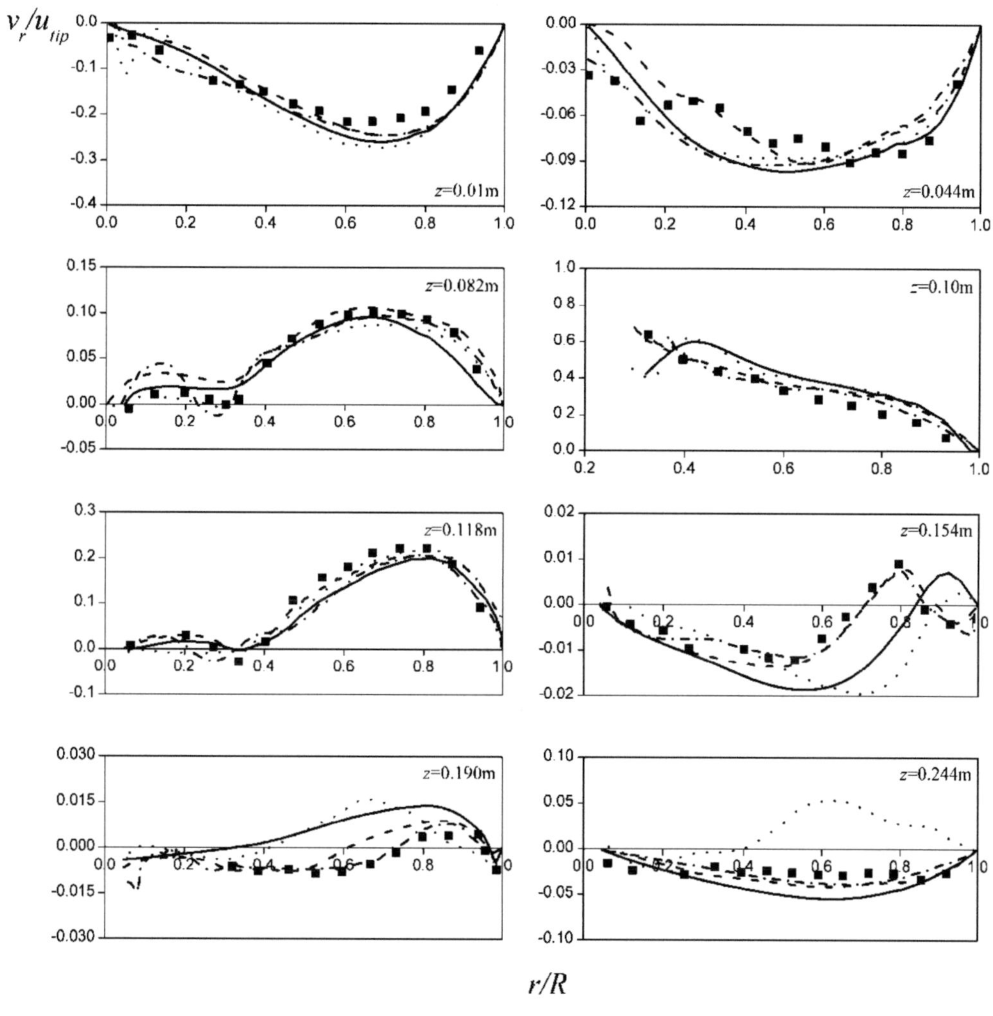

Figure 5. Comparison of the results predicted by different impeller approaches and experimental data of the dimensionless mean radial velocity in the bulk region: experimental data (Murthy and Joshi, 2008) standard k-ε momentum source term approach predictions, RSM momentum source term approach predictions, standard k-ε SM predictions (Murthy and Joshi, 2008), RSM SM predictions (Murthy and Joshi, 2008).

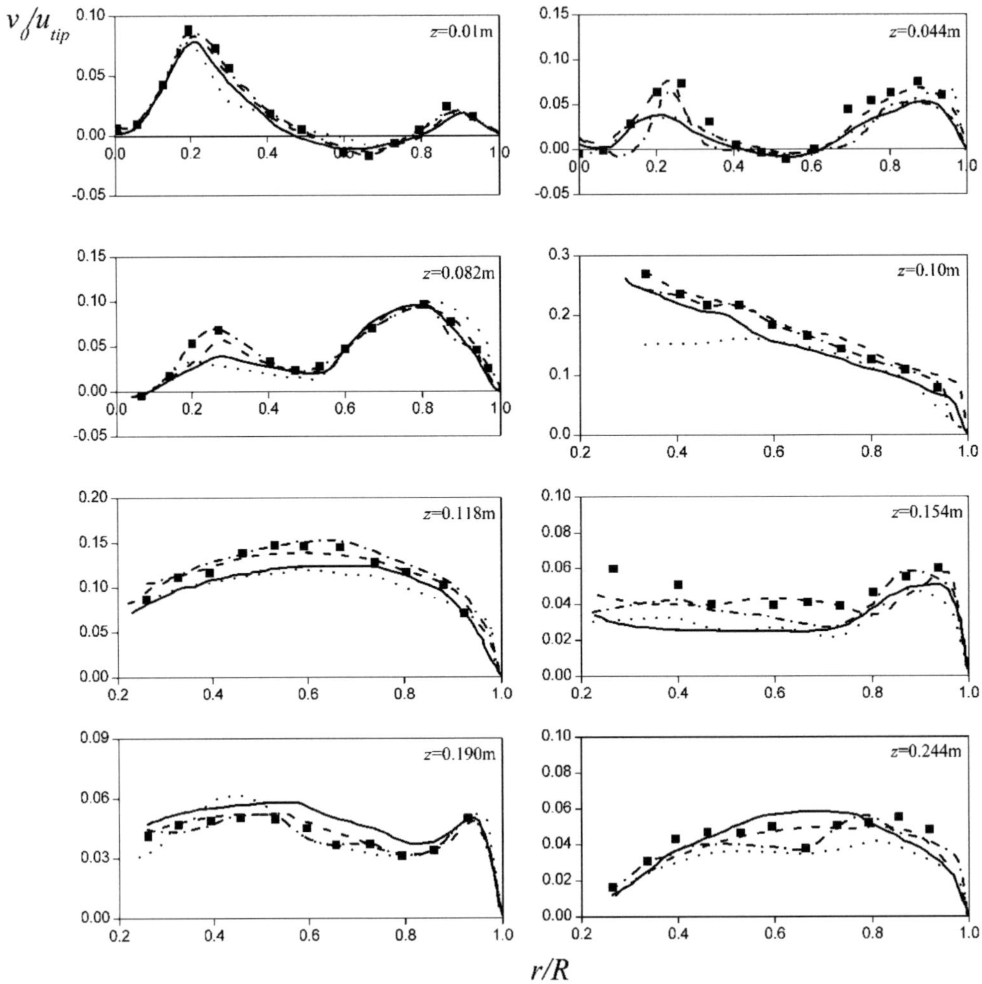

Figure 6. Comparison of the results predicted by different impeller approaches and experimental data of the dimensionless mean tangential velocity in the bulk region: experimental data (Murthy and Joshi, 2008) standard *k*-ε momentum source term approach predictions, RSM momentum source term approach predictions, standard *k*-ε SM predictions (Murthy and Joshi, 2008), RSM SM predictions (Murthy and Joshi, 2008).

Fig.6 illustrates the radial profiles of the tangential velocity component. Yianneskis et al [40] pointed out that the baffles reduce the vessel cross-section, which results in higher values for tangential velocity and a reduced pressure, thus generating reverse flows. It may be the reason that negative velocities exist at the levels of z = 0.01 m and z = 0.044 m. It can be seen that the computational results of momentum source term approach and SM with standard *k*-ε model are similar, and both of them predict the reverse flows. They agree rather well with the experimental data at the lower levels, whereas deviate at the upper levels. For RSM simulations, the prediction accuracy of two modeling methods have been improved, and their results are in good agreement with the experimental results.

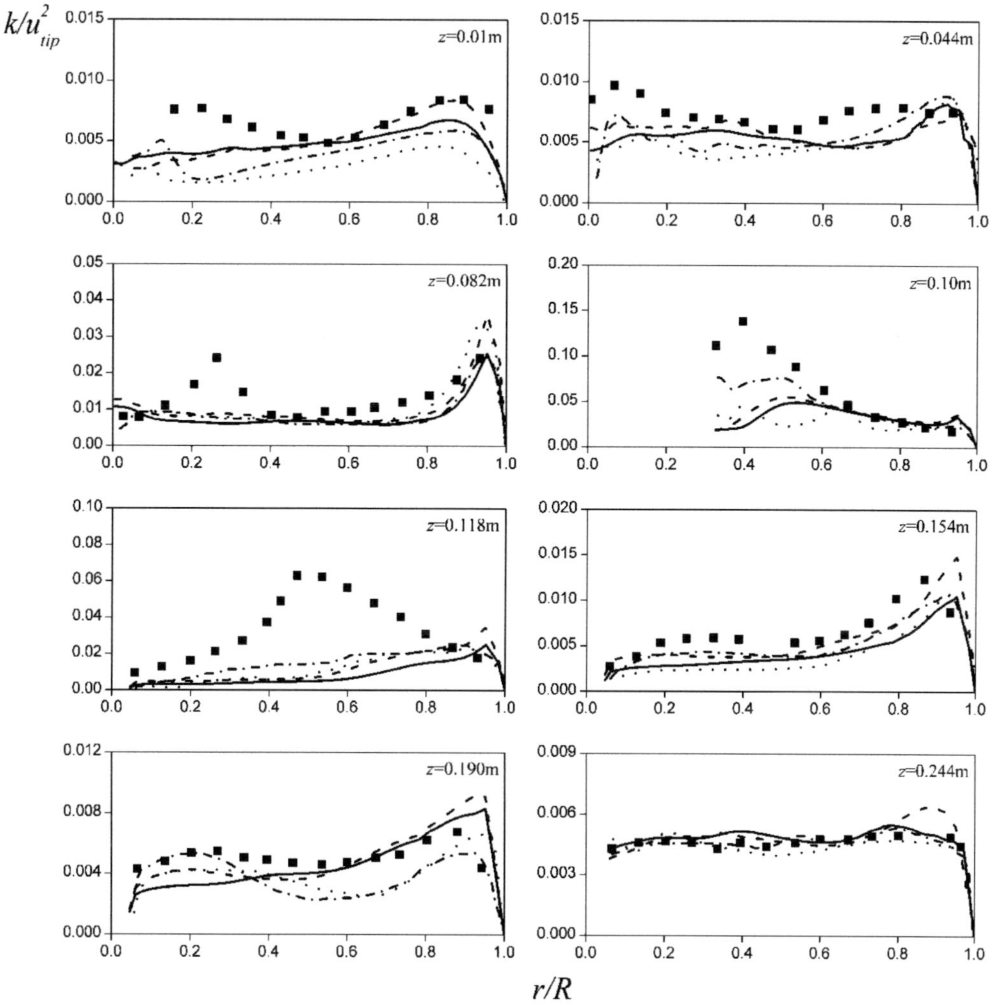

Figure 7. Comparison of the results predicted by different impeller approaches and experimental data of the dimensionless mean turbulent kinetic energy in the bulk region: experimental data (Murthy and Joshi, 2008) standard *k-ε* momentum source term approach predictions, RSM momentum source term approach predictions, standard *k-ε* SM predictions (Murthy and Joshi, 2008), RSM SM predictions (Murthy and Joshi, 2008).

Fig.7 shows the numerical comparison of turbulent kinetic energy in various positions. At z = 0.082m, it can be noticed that there are two maxima of turbulent kinetic energy, similar to the case of tangential velocity (Fig.6). They are generated by the interactions between fluid and blades a, fluid and wall, respectively. Overall, the CFD results which predicted by standard *k-ε* model are not satisfying, significantly underpredicting the turbulent kinetic energy at almost all positions. However, the results of momentum source term approach and SM are similar. Although RSM predictions of momentum source term approach and SM have reduced the errors, both still underestimate turbulent kinetic energy, and their results are similar. Due to unsteady and complex nature of flow characteristics in the impeller region and the continuous conversion of kinetic energy [37], the k-ε model and RSM both fail to capture the transfer details of kinetic energy, so the momentum source term approach and SM model underpredict the turbulent kinetic energy with similar deviations when the k-ε and RSM turbulent model are applied.

From the comparisons above, it can be found that predictions of momentum source term approach with RSM turbulent model are in good agreement with the experimental data as well as SM model, although both SM and our model have some deviations in prediction of the upper part for the radial and tangential velocity. The models combined with k-ε turbulent model have less prediction accuracy, Murthy and Joshi [22] attributed these numerical errors to the overestimation of the eddy viscosity of the k-ε model. It is suggested that the standard k–ε model performs well when the flow is unidirectional that is with less swirl and weak recirculation [11, 22].

3.3. Velocity vectors

Fig.8 shows velocity vectors of the vertical plane in the middle of two baffles. It can be seen that after exerting on the momentum source, fluid velocity in the impeller region is very high, and flow gradients are large.

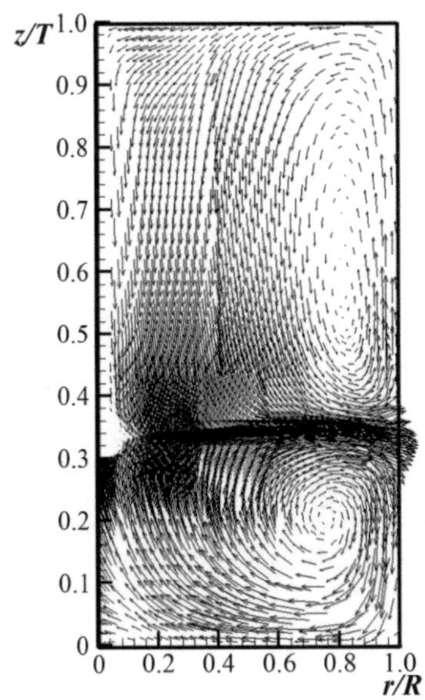

Figure 8. Velocity vectors of stirred tank in the middle plane.

As the high speed fluid jets outward, initially almost not affected by the surrounding fluid, the velocity contours are dense. The flow impinges on the tank wall, splits up into two parts and changes the direction. The split water flows at last return to impeller region and accelerated again, repeating above-mentioned process which generates two circulation loops of different directions in the upper and lower part of the tank, respectively. It can be observed that the circulation loop ranges above the z/T value of 0.9, which is consistent with the simulations by Ng et al. [24] (1998) (Re = 40,000) and Yapici et al. [41] (2008) (Re = 60,236). Reynolds numbers of previous and present studies can ensure that the flow in a stirred vessel is fully developed turbulence, so the circulation pattern predictions can be considered properly.

Fig.9 shows the flow field of stirred tank near the impeller tip. It can be noticed that velocity distributions are not symmetrical about the impeller centre plate (z = 0.1 m), but shift upwards slightly. This result is in agreement with the previous experimental works, that the impeller is not symmetrically located, the top of the tank is a free surface, and the hub is present on the top side of the impeller [38, 42-43]. In this study, this detail has been successfully captured in the velocity field near impeller tip, further indicating that momentum source term approach prediction is in good agreement with experimental results.

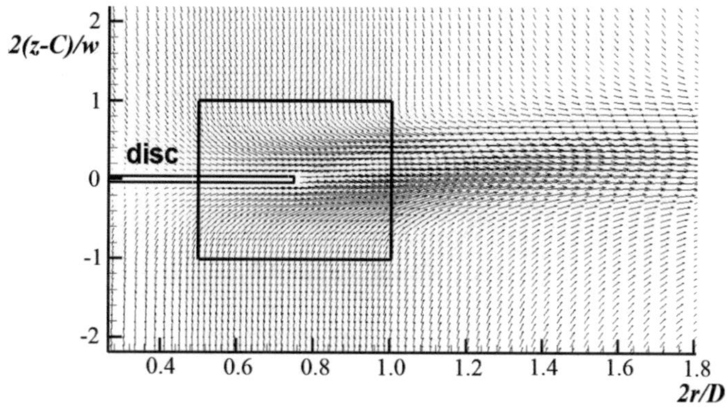

Figure 9. Velocity vectors in the middle plane near impeller tip.

3.4. Comparison of power numbers

Table 2 shows the comparison of predicted power numbers with the experimental value. Here, the experimental power numbers reported by Rushton et al. [44-45], Murthy and Joshi [22] are applied in the reference. The quantities obtained by SM model through integral ε based approach are considerably lower than the reported experimental results. Similar results have been observed in earlier studies [29, 46-47]. These deviations are usually attributed to underestimation of the turbulent quantities associated to the k–ε model [46, 48]. The power number predictions of integral power based approach in the momentum source term approach with standard k–ε model and RSM are 5.72 and 5.64, respectively, both between the experimental data 5.1 and 6.07, so they are better than those of SM.

Fig.10 shows a log-log plot of the experimentally obtained and predicted power numbers for nine kinds of Reynolds numbers which cover laminar and turbulent flow regimes. It can be observed that at lower Reynolds numbers the power numbers predicted by momentum source term approach are in good agreement with the experimental data of Rushton et al. [44].

Methods	N_P	Error/%
Experimental value (Rushton et al., 1950)	6.07	/
Experimental value (Murthy and Joshi, 2008)	5.1	/
SM torque based approach (Murthy and Joshi, 2008)	4.9	12.26
SM integral ε based approach (Murthy and Joshi, 2008)	3.9	30.17
MRF torque based approach (Deglon and Meyer, 2006)	5.40	3.31
Standard *k*-ε integral power based approach	5.72	2.42
RSM integral power based approach	5.64	0.98

Table 2. Power number predictions for different approaches at *Re*=41,467.

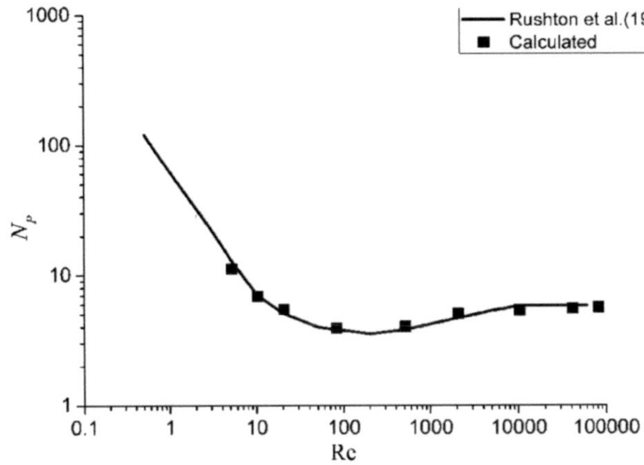

Figure 10. Comparison of experimental and predicted power numbers.

3.5. Computational speed

The simulations were carried out on a 100 node AMD64 cluster, each node including 2 four-core processors with 2.26 GHz clock speed and 2 GB memory. 80 processors were applied in all the computations. Present study compared the computational speeds of momentum source term approach, MRF and SM, and the required expenses are shown in Table 3. It can be seen that momentum source term approach and MRF using steady state simulations need less computational requirements than SM, and the computational time of momentum source term approach is the least.

Approach	Momentum source term	MRF	SM
Standard *k*-ε	48	72	340
RSM	60	86	410

Table 3. Comparison of the computational time required by different approaches, h.

http://dx.doi.org/10.5772/51754

4. Conclusions

An equation to predict the momentum source term is proposed without the help of experimental data in the paper. So the momentum source term approach for CFD prediction of the impeller propelling action has been developed as a tool with predictive capacity. The prediction results of the approach have been compared with the experimental data, MRF and SM model predictions in the literatures. The following conclusions can be drawn from the present work:

1. For the plate blade of the Rushton turbine stirred vessel, the tangential momentum source added by blade is proposed to be calculated by:

in which u is the linear velocity of the area element on the blade surface dS; v_{θ} is the fluid tangential velocity rotating with the impeller before the fluid was propelled by impeller; dS is cross-section area of the interface between fluid and the impeller blade. The radial friction force of the impeller blade is proposed to be calculated approximately (Wang et al., 2002) as following:

In which f is the friction force in the computational cells, v_r is the radial velocity of the fluid.

2. The numerical comparisons of flow field show that the momentum source term approach predictions are in good agreement with the experimental data. It has been also found that the prediction accuracy of momentum source term approach is better than MRF and similar to SM, whereas the computational time of momentum source term approach is the least of the three.

Acknowledgements

This work is supported by the Ministry of Science and Technology of the People's Republic of China (grants No. 2006BAC19B02). The authors would like to thank professor Taohong Ye for helpful discussion, the supercomputing center of University of Science and Technology of China for their help in computing.

Author details

Weidong Huang* and Kun Li

*Address all correspondence to: huangwd@ustc.edu.cn

Department of Geochemistry and Environmental Science, University of Science and Technology of China, P. R. China

References

[1] Harvey, P. S., & Greaves, M. (1982). Turbulent-Flow in an Agitated Vessel.1. A Predictive Model. *Transactions of the Institution of Chemical Engineers*, 60(4), 195-200.

[2] Harvey, P. S., & Greaves, M. (1982). Turbulent-Flow in an Agitated Vessel.2. Numerical-Solution and Model Predictions. *Transactions of the Institution of Chemical Engineers*, 60(4), 201-210.

[3] Ranade, V. V., & Joshi, J. B. (1990). Flow generated by a disc turbine Part II: mathematical modelling and comparison with experimental data. *Institution of Chemical Engineering* [68A], 34-43.

[4] Kresta, S. M., & Wood, P. E. (1991). Prediction of the three-dimensional turbulent flow in stirred tanks. *Aiche Journal*, 37, 448-460.

[5] Luo, J. Y., Issa, R. I., & Gosman, A. D. (1994). Prediction of impeller-induced flows in mixing vessels using multiple frames of reference. Cambridge, UK. *in Proceeding of 8th Europe Conference on Mixing*, 155-162.

[6] Luo, J. Y., Gosman, A. D., Issa, R. I., et al. (1993). Full Flow-Field Computation of Mixing in Baffled Stirred Vessels. *Chemical Engineering Research & Design*, 71(A3), 342-344.

[7] Murthy, J. Y., Mathur, S. R., & Choudhary, D. (1994). CFD simulation of flows in stirred tank rectors using a sliding mesh technique. *in Proceeding of 8th Europe Conference on Mixing, Cambridge, UK*, 155-162.

[8] Mostek, M., Kukukova, A., Jahoda, M., et al. (2005). Comparison of different techniques for modelling of flow field and homogenization in stirred vessels. *Chemical Papers*, 59(6A), 380-385.

[9] Perng, C. Y., & Murthy, J. Y. (1994). A moving deforming mesh technique for simulation of flow in mixing tanks. Cambridge, UK. *in Proceeding of 8th Europe Conference on Mixing*, 37-39.

[10] Joshi, J. B., Nere, N. K., Rane, C. V., et al. (2011). Cfd Simulation of Stirred Tanks: Comparison of Turbulence Models. Part II: Axial Flow Impellers, Multiple Impellers and Multiphase Dispersions. *Canadian Journal of Chemical Engineering*, 89(4), 754-816.

[11] Joshi, J. B., Nere, N. K., Rane, C. V., et al. (2011). Cfd Simulation of Stirred Tanks: Comparison of Turbulence Models. Part I: Radial Flow Impellers. *Canadian Journal of Chemical Engineering*, 89(1), 23-82.

[12] Pericleous, K. A., & Patel, M. K. (1987). The modelling of tangential and axial agitators in chemical reactors. *Physico-chimie Chemical Hydrodynamics*, 8, 105-123.

[13] Dhainaut, M., Tetlie, P., & Bech, K. (2005). Modeling and experimental study of a stirred tank reactor. *International Journal of Chemical Reactor Engineering*, 3.

[14] Xu, Y., & Mc Grath, G. (1996). CFD predictions in stirred tank flows. *Chemical Engineering Research and Design.*, 74, 471-475.

[15] Patwardhan, A. W. (2001). Prediction of flow characteristics and energy balance for a variety of downflow impellers. *Industrial & Engineering Chemistry Research*, 40(17), 3806-3816.

[16] Revstedt, J., Fuchs, L., & Tragardh, C. (1998). Large eddy simulations of the turbulent flow in a stirred reactor. *Chemical Engineering Science*, 53(24), 4041-4053.

[17] Schneekluth, H. (1988). *Hydromechanik zum Schiffsentwurf : Vorlesungen. 3., verb. und erw. Aufl. ed., Herford: Koehler. ix*, 1076.

[18] Jiang, C. Y. (2007). *Study of several CFD models in the oxidation ditch system*, University of Science and Technology of China, Hefei, China.

[19] Jiang, C. Y., Huang, W. D., Wang, G., et al. (2010). Numerical computation of flow fields in an oxidation ditch by computational fluid dynamics model. *Environmental Science & Technology*, 33, 135-140.

[20] Gou, Q. (2008). *Numerical simulation of the propellers in oxidation ditch using Computational fluid dynamics models in Department of Earth and Space Science*, University of Science and Technology of China, Hefei.

[21] Li, B., Zhang, Q. W., Hong, H. S., & You, T. (2009). Several factors of CFD numerical simulation in stirred tank. *Chemical Industry and Engineering Progress*, 28, 7-12.

[22] Joshi, J. B., & Murthy, B. N. (2008). Assessment of standard k-epsilon, RSM and LES turbulence models in a baffled stirred vessel agitated by various impeller designs. *Chemical Engineering Science*, 63(22), 5468-5495.

[23] Deglon, D. A., & Meyer, C. J. (2006). CFD modelling of stirred tanks: Numerical considerations. *Minerals Engineering*, 19(10), 1059-1068.

[24] Ng, K., Fentiman, N. J., Lee, K. C., et al. (1998). Assessment of sliding mesh CFD predictions and LDA measurements of the flow in a tank stirred by a Rushton impeller. *Chemical Engineering Research & Design*, 76(A6), 737-747.

[25] Ding, Z. R. (2002). *Hydrodynamics*, Beijing, China, Higher Education Press.

[26] Schneekluth, H. (1997). *Hydromechanik zum schiffsentwurf*, Herford, Koehler Book Company, 3, In Chinese, translated by Xian P.L., Shanghai, China: Publishing company of Shanghai Jiao Tong University, 1988.

[27] Wang, Y. D., Luo, G. S., & Liu, Q. (2002). *Principles of transport processes in chemical engineering*, Beijing, China, Tsinghua University Press.

[28] Wechsler, K., Breuer, M., & Durst, F. (1999). Steady and unsteady computations of turbulent flows induced by a 4/45 degrees pitched-blade impeller. *Journal of Fluids Engineering*, 121, 318-329.

[29] Montante, G, Coroneo, M, Paglianti, A, et al. (2011). CFD prediction of fluid flow and mixing in stirred tanks: Numerical issues about the RANS simulations. *Computers & Chemical Engineering*, 35(10), 1959-1968.

[30] Aubin, J., Fletcher, D. F., & Xuereb, C. (2004). Modeling turbulent flow in stirred tanks with CFD: the influence of the modeling approach, turbulence model and numerical scheme. *Experimental Thermal and Fluid Science*, 28(5), 431-445.

[31] Launder, B. E., & Spalding, D. B. (1974). Numerical computation of turbulent flows. *Computer Methods in Applied Mechanics and Engineering*, 3, 269-289.

[32] Speziale, C. G., Sarkar, S., & Gatski, T. B. (1991). Modelling the pressure-strain correlation of turbulence: an invariant dynamical systems approach. *Journal of Fluid Mechanics*, 227, 245-272.

[33] Brucato, A. M. C. F. G., et al. (1998). Numerical prediction of flows in baffled stirred vessels: a comparison of alternative modelling approaches. *Chemical Engineering Science*, 53, 3653-3684.

[34] Bartels, C., Breuer, M., Wechsler, K., et al. (2002). Computational fluid dynamics applications on parallel-vector computers: computations of stirred vessel flows. *Computers & Fluids*, 31(1), 69-97.

[35] Xuereb, C., & Bertrand, J. (1996). D hydrodynamics in a tank stirred by a double-propeller system and filled with a liquid having evolving rheological properties. *Chemical Engineering Science*, 51(10), 1725-1734.

[36] Shekhar, S. M., & Jayanti, S. (2002). CFD study of power and mixing time for paddle mixing in unbaffled vessels. *Chemical Engineering Research & Design*, 80(A5), 482-498.

[37] Line, A., & Escudie, R. (2003). Experimental analysis of hydrodynamics in a radially agitated tank. *Aiche Journal*, 49(3), 585-603.

[38] Wu, H, & Patterson, G.K. (1989). Laser-doppler measurements of turbulent flow parameters in a stirred mixer. *Chemical Engineering Science*, 44, 2207-2221.

[39] Murthy, B. N., & Joshi, J. B. (2008). Assessment of standard k-ε, RSM and LES turbulence models in a baffled stirred vessel agitated by various impeller designs. *Chemical Engineering Science*, 63, 5468-5495.

[40] Yianneskis, M., Popiolek, Z., & Whitelaw, J. H. (1987). An experimental study of the steady and unsteady flow characteristics of stirred reactors. *Journal of Fluid mechanics*, 175, 537-555.

[41] Yapici, K., Karasozen, B., Schafer, M., et al. (2008). Numerical investigation of the effect of the Rushton type turbine design factors on agitated tank flow characteristics. *Chemical Engineering and Processing*, 47, 1340-1349.

[42] Stoots, C. M., & Calabresc, R. V. (1995). Mean velocity field relative to a rushton turbine blade. *Aiche Journal*, 41, 1-11.

[43] Derksen, J. J., Doelman, M. S., & Van den Akker, H. E. A. (1999). Three-dimensional LDA measurements in the impeller region of a turbulently stirred tank. *Experiments in Fluids*, 27(6), 522-532.

[44] Rushton, J. H., Costich, E. W., & Everett, H. J. (1950). Power Characteristics of Mixing Impellers.1. *Chemical Engineering Progress*, 46(8), 395-404.

[45] Rushton, J. H., Costich, E. W., & Everett, H. J. (1950). Power Characteristics of Mixing Impellers.2. *Chemical Engineering Progress*, 46(9), 467-476.

[46] Kukukova, A., Mostek, M., Jahoda, M., et al. (2005). CFD prediction of flow and homogenization in a stirred vessel: Part I vessel with one and two impellers. *Chemical Engineering & Technology*, 28(10), 1125-1133.

[47] Uludag, Y., Yapici, K., Karasozen, B., et al. (2008). Numerical investigation of the effect of the Rushton type turbine design factors on agitated tank flow characteristics. *Chemical Engineering and Processing*, 47(8), 1346-1355.

[48] Sommerfeld, M., & Decker, S. (2004). State of the art and future trends in CFD simulation of stirred vessel hydrodynamics. *Chemical Engineering & Technology*, 27(3), 215-224.

Chapter 6

Hydrodynamic and Heat Transfer Simulation of Fluidized Bed Using CFD

Osama Sayed Abd El Kawi Ali

Additional information is available at the end of the chapter

http://dx.doi.org/10.5772/52072

1. Introduction

The nuclear energy is suffering from the lack of public acceptance everywhere mainly due to the issues relating to reactor safety, economy and nuclear waste. The Fluidized Bed Nuclear Reactor (FBNR) concept has addressed these issues and tried to resolve such problems. The FBNR is small, modular and simple in design contributing to the economy of the reactor. It has inherent safety and passive cooling characteristics. Its spent fuel being small spherical elements may not be considered nuclear waste, and can be directly used as a source of radiation for applications in industry and agriculture resulting in reduced environmental impact [1].

With the increase of computational power, numerical simulation becomes an additional tool to predict the fluid dynamics and the heat transfer mechanism in multiphase flow. A numerical hydrodynamic and heat transfer model has been developed to simulate the gas fluidized bed. All of CFD, in one form or another, is based on the fundamental governing equations of fluid dynamics (continuity, momentum and energy equations). These equations speak physics. They are mathematical statements of three fundamental physical principals upon which all fluid dynamics is based: mass is conserved, Newton's second law and energy is conserved [2].

This chapter aims to study a mathematical modeling and numerical simulation of the hydrodynamics and heat transfer processes in a two-dimensional gas fluidized bed with a vertical uniform gas velocity at the inlet. The velocity, volume fraction, temperature distribution for gas phase and particle phase are calculated. Also, gas pressure and a prediction of the average heat transfer coefficient are also studied.

Such a simulation technique allows performance evaluation for different bed input parameters, and can evolve into a tool for optimized design of fluidized beds for different industrial use.

The numerical setup consists of a two dimensional fluidized bed filled with particles.The cold gas enters to the bed to cool the hot particles. Based on conservation equations for both phases it is possible to predict particles and gas volume fractions, velocity distributions (for gas and particles), temperature distribution, heat transfer coefficient as well as gas pressure field.

2. General assumptions for the mathematical model

Fluidized beds are categorized as multiphase flow problems. There are currently two approaches to model multiphase flow problems as discussed in chapter two. The best overall balance between computational time and accuracy seems to be achieved by implementing an Eulerian-Eulerian approach. The following assumptions are introduced into the present analysis:

1. The bed is two-dimensional.
2. Eulerian-Eulerian approach is applied.
3. The gas has constant physical properties.
4. Uniform fluidization.
5. Constant input fluid flux.
6. There is no mass transfer or chemical reaction between the two phases.
7. All particles are spherical in shape with the same diameter,d_p.
8. The expanded bed region is considered in the analysis in addition to the out flowing gas, i.e. suspension and free board regions.
9. Viscous heat dissipation in the energy equation is negligible in comparison with conduction and convection.

3. Governing equations

Due to the high particle concentration in fluidized beds the particles interactions cannot be neglected. In fact the solid phase has similar properties as continuous fluid. Therefore, the Eulerian approach is an efficient method for the numerical simulation of fluidized beds.

A hydrodynamic and thermal model for the fluidized bed is developed based on schematic diagram shown in Figure (1). The principles of conservation of mass, momentum, and energy are used in the hydrodynamic and thermal models of fluidization. The general mass conservation equations and the separate phase momentum equations and energy equations (for each phase) for fluid–solids, nonreactive transient and two-phase flow will be discussed in the following sections.

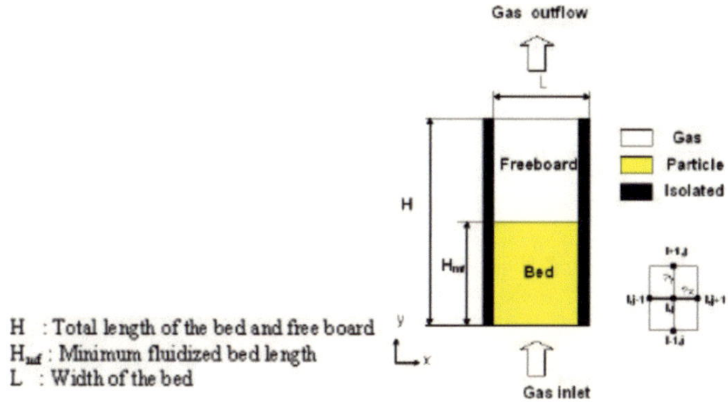

Figure 1. The schematic diagram of the present work model

3.1. Continuity equation

Particle phase:

$$\frac{\partial \varepsilon_s}{\partial t}+\frac{\partial}{\partial x}(\varepsilon_s u_s)+\frac{\partial}{\partial y}(\varepsilon_s v_s)=0 \tag{1}$$

Gas phase :

$$\frac{\partial \varepsilon_g}{\partial t}+\frac{\partial}{\partial x}(\varepsilon_g u_g)+\frac{\partial}{\partial y}(\varepsilon_g v_g)=0 \tag{2}$$

3.2. Volume fraction constraint

$$\varepsilon_g+\varepsilon_s=1.0 \tag{3}$$

3.3. Momentum equation

Particle phase:

$$\rho_s\varepsilon_s\frac{\partial u_s}{\partial t}+\rho_s\varepsilon_s\frac{\partial}{\partial x}(u_s u_s)+\rho_s\varepsilon_s\frac{\partial}{\partial y}(v_s u_s)=F_{sx}\ \text{“x -direction“} \tag{4}$$

$$\rho_s\varepsilon_s\frac{\partial v_s}{\partial t}+\rho_s\varepsilon_s\frac{\partial}{\partial x}(u_s v_s)+\rho_s\varepsilon_s\frac{\partial}{\partial y}(v_s v_s)=F_{sy}\ \text{“y -direction“} \tag{5}$$

The total force acting on particle phase is the sum of the net primary force and the force resulting from particle phase elasticity. The x and y components of forces acting on particle phase are as following:

$$F_{sy} = C_d \frac{3\varepsilon_s \rho_g (v_g - v_s)\left|v_g - v_s\right|}{4d_p}(1-\varepsilon_s)^{-1.8} - \varepsilon_s \rho_s g$$
$$-\varepsilon_s \frac{\partial p}{\partial y} - \frac{\partial \varepsilon_s}{\partial y}(3.2 g d_p \varepsilon_s (\rho_s - \rho_g)) \tag{6}$$

$$F_{sx} = C_d \frac{3\varepsilon_s \rho_g (u_g - u_s)\left|u_g - u_s\right|}{4d_p}(1-\varepsilon_s)^{-1.8} - \varepsilon_s \frac{\partial p}{\partial x} \tag{7}$$

Gas phase :

$$\rho_g \varepsilon_g \frac{\partial u_g}{\partial t} + \rho_g \varepsilon_g \frac{\partial}{\partial x}(u_g u_g) + \rho_g \varepsilon_g \frac{\partial}{\partial y}(v_g u_g) = F_{gx} \quad \text{"x -direction"} \tag{8}$$

$$\rho_g \varepsilon_g \frac{\partial v_g}{\partial t} + \rho_g \varepsilon_g \frac{\partial}{\partial x}(u_g v_g) + \rho_g \varepsilon_g \frac{\partial}{\partial y}(v_g v_g) = F_{gy} \quad \text{"y -direction"} \tag{9}$$

The fluid phase forces are readily obtained from the particle phase relations for fluid – particle interaction (drag and pressure gradient force), which act in the opposite direction on the fluid, together with gravity. The x and y components of forces acting on gas phase are as following:

$$F_{gy} = -C_d \frac{3\varepsilon_s \rho_g (v_g - v_s)\left|v_g - v_s\right|}{4d_p}(1-\varepsilon_s)^{-1.8} - \varepsilon_g \rho_s g - \varepsilon_g \frac{\partial p}{\partial y} \tag{10}$$

$$F_{gx} = -F_{sx} = -C_d \frac{3\varepsilon_s \rho_g (u_g - u_s)\left|u_g - u_s\right|}{4d_p}(1-\varepsilon_s)^{-1.8} + \varepsilon_s \frac{\partial p}{\partial x} \tag{11}$$

3.3.1. Relation between fluid and particle velocities

We assume that both particles and fluid are regarded as being incompressible. This was justified on the basis that only a gas phase is going to exhibit any significant compressibility, and the orders of magnitude differences in particle and fluid density for gas fluidization render quite insignificant the small change in gas density resulting from compression. This assumption led to the relation linking fluid and particle phase velocities at all location. By applying the overall mass balance, which is obtained by summing equations (1) and (2):

$$\frac{\partial}{\partial x}(\varepsilon_s u_s + \varepsilon_g u_g) + \frac{\partial}{\partial y}(\varepsilon_s v_s + \varepsilon_g v_g) = 0 \tag{12}$$

Equation (12) shows that the total flux (fluid plus particles) in x- direction and y-direction remains constant, equal to that of fluid entering the bed U_{gin}, V_{gin}. This result is a simple consequence of the particles and fluid being considered incompressible :

$$V_{gin} = \varepsilon_g v_g + \varepsilon_s v_s \tag{13}$$

$$U_{gin} = \varepsilon_g u_g + \varepsilon_s u_s \tag{14}$$

Equations (13) and (14) enable the fluid velocity variables to be expressed in terms of the particle velocity at all points in the bed.

3.3.2. Combined momentum equation

In this section the combined momentum equation is produced by combining the fluid and particle momentum equations (4), (5), (8) and (9) by elimination of the fluid pressure gradient, which appears in them. This yields the combined momentum equation:

x- direction:

[divide equation (4-4) by ε_s] + [divide equation (4-8) by ε_g] which give us :

$$\rho_s\left[\frac{\partial u_s}{\partial t} + \frac{\partial}{\partial x}(u_s u_s) + \frac{\partial}{\partial y}(v_s u_s)\right] + \rho_g\left[\frac{\partial u_s}{\partial t} + \frac{\partial}{\partial x}(u_g u_g) + \frac{\partial}{\partial y}(v_g u_g)\right] = 0 \tag{15}$$

y- direction:

[divide equation (4-3) by ε_s] – [divide equation (4-7) by ε_g] which give us :

$$\rho_s\left[\frac{\partial v_s}{\partial t} + \frac{\partial}{\partial x}(u_s v_s) + \frac{\partial}{\partial y}(v_s v_s)\right]$$

$$-\rho_g\left[\frac{\partial v_g}{\partial t} + \frac{\partial}{\partial x}(u_g v_g) + \frac{\partial}{\partial y}(v_g v_g)\right] = \frac{F_{sy}}{\varepsilon_s} - \frac{F_{gy}}{\varepsilon_g} \tag{16}$$

Equations (15) and (16) with the continuity equation for the two phase (1) and (2),now define the two phase system without account of fluid pressure variation.

3.3.3. Drag coefficient

An important constitutive relation in any multiphase flow model is the formula for the fluid-particle drag coefficient, which is may be expressed by the empirical Dallavalle relation as reported in [3] :

$$C_d = \left(0.63 + \frac{4.8}{\mathrm{Re}^{0.5}}\right)^2 \tag{17}$$

The particle Reynolds number, Re, based on particle diameter is given by :

$$\mathrm{Re} = \frac{\varepsilon_g \rho_g d_p \left|\overrightarrow{U_r}\right|}{\mu_g} \tag{18}$$

3.3.4. Gas pressure drop

Figure (2) shows the relation between the total pressure drop across the bed "ΔP_B" and the input gas velocity [3].where:

$$\Delta P_B = \left(\rho_g \varepsilon_{g,mf} + \rho_s\left(1 - \varepsilon_{g,mf}\right)\right) g H_{mf} \tag{19}$$

The gas pressure at the entrance of the fluidized bed can be calculated from the following equation:

$$P_{g,in} = P_{atm} + \rho_g g\left(H - H_{mf}\right) + \left(\rho_g \varepsilon_{g,mf} + \rho_s\left(1 - \varepsilon_{g,mf}\right)\right) g H_{mf} \tag{20}$$

and the pressure drop at any position and head "h" can be calculated from:

$$\Delta P_g = \rho_{sus} g h \tag{21}$$

Where ρ_{sus} is the suspension density which calculated from:

$$\rho_{sus} = \rho_s \varepsilon_s + \rho_g \varepsilon_g \tag{22}$$

http://dx.doi.org/10.5772/52072

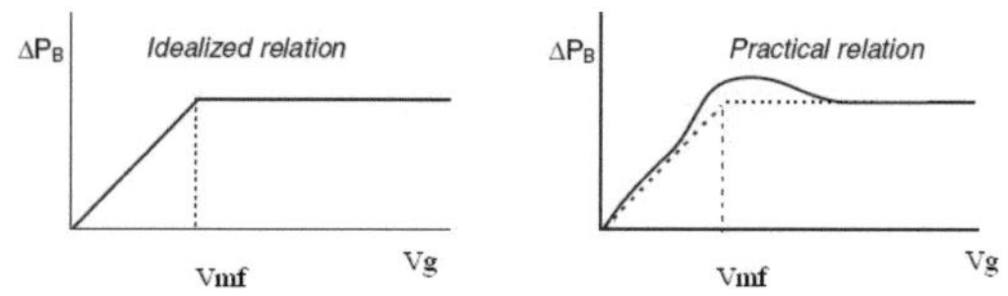

Figure 2. Total pressure drop in fluidized bed

3.4. Energy equation

Particle phase:

$$\rho_s C_{ps} \frac{\partial \varepsilon_s T_s}{\partial t} + \rho_s C_{ps} \frac{\partial}{\partial x}(\varepsilon_s u_s T_s) + \rho_s C_{ps} \frac{\partial}{\partial y}(\varepsilon_s v_s T_s) = \frac{\partial}{\partial x}(\varepsilon_s k_s \frac{\partial T_s}{\partial x}) + \frac{\partial}{\partial y}(\varepsilon_s k_s \frac{\partial T_s}{\partial y}) + h_v(T_g - T_s) + \varepsilon_s \dot{q} \tag{23}$$

GAS phase:

$$\rho_g C_{pg} \frac{\partial \varepsilon_g T_g}{\partial t} + \rho_g C_{pg} \frac{\partial}{\partial x}(\varepsilon_g u_g T_g) + \rho_g C_{pg} \frac{\partial}{\partial y}(\varepsilon_g v_g T_g) = \frac{\partial}{\partial x}(\varepsilon_g k_g \frac{\partial T_g}{\partial x}) + \frac{\partial}{\partial y}(\varepsilon_g k_g \frac{\partial T_g}{\partial y}) + h_v(T_s - T_g) \tag{24}$$

3.4.1. Thermal conductivity values (k_g and k_s)

The thermal conductivities of the fluid phase and the solid phase (k_g and k_s) in the two fluid model formulation should be interpreted as effective transport coefficients which means that the corresponding microscopic (or absolute) coefficients $k_{g,o}$ and $k_{s,o}$ cannot be used. It can be represented in general as:

$$k_g = k_g(k_{g,o}, k_{s,o}, \varepsilon_g, \text{particle geometry}) \tag{25}$$

$$k_s = k_s(k_{g,o}, k_{s,o}, \varepsilon_g, \text{particle geometry}) \tag{26}$$

However, such a general formulation is not yet available for fluidized beds and approximate constitutive equations have to be used. These approximate equations have been obtained on modeling of the effective thermal conductivity k_b in packed beds. According to their model, the radial bed conductivity k_b consists of a contribution $k_{b,g}$ due to the fluid phase only and a contribution due to a combination of the fluid phase and the solid phase[4].

$$k_b = k_{b,g} + k_{b,s} \tag{27}$$

where :

$$k_{bg} = \left(1 - \sqrt{1 - \varepsilon_g}\right) k_{go} \tag{28}$$

$$k_{bs} = \sqrt{1 - \varepsilon_g} \left(\omega A + (1 - \omega)\Gamma\right) k_{go} \tag{29}$$

$$\Gamma = \frac{2}{\left(1 - \frac{B}{A}\right)} \left[\frac{A - 1}{\left(1 - \frac{B}{A}\right)^2} \frac{B}{A} Ln\left(\frac{A}{B}\right) - \frac{B - 1}{\left(1 - \frac{B}{A}\right)} - \frac{1}{2}(B + 1) \right] \tag{30}$$

$$B = 1.25 \left(\frac{1 - \varepsilon_g}{\varepsilon_g} \right)^{\frac{10}{9}} \tag{31}$$

$$A = \frac{k_{s,o}}{k_{g,o}} \tag{32}$$

$$\omega = 7.26X10^{-3} \tag{33}$$

Thus the thermal conductivities of the fluid phase and the solid phase then are given by :

$$k_g = \frac{k_{b,g}}{\varepsilon_g} \tag{34}$$

$$k_s = \frac{k_{b,s}}{\varepsilon_s} \tag{35}$$

3.4.2. Interphase volumetric heat transfer coefficient h_v

The heat transfer coefficient is modelled using a correlation by Gunn as reported [5]. This correlation is applicable for gas voidage in the range of 0.35 to 1 and for Reynolds numbers up to Re = 10^5, and gives the Nusselt number:

$$Nu = \frac{h_{gp} d_p}{k_{g,o}} \tag{36}$$

$$Nu = \left(7 - 10\varepsilon_g + 5\varepsilon_g^{2}\right)\left(1 + 0.7\mathrm{Re}_p^{0.2}\mathrm{Pr}^{\frac{1}{3}}\right)$$

$$+\left(1.33 - 2.40\varepsilon_g + 1.20\varepsilon_g^2\right)\mathrm{Re}_p^{0.7}\,\mathrm{Pr}^{\frac{1}{3}} \tag{37}$$

where Reynolds number is defined by equation (18).The Prandlt number is defined by;

$$\mathrm{Pr} = \frac{C_{p,g}\mu_g}{k_{g,o}} \tag{38}$$

,and the overall heat transfer coefficient is evaluated from;

$$h_v = \frac{6\left(1 - \varepsilon_g\right)h_{gp}}{d_p} \tag{39}$$

4. Boundary and initial conditions

The system of conservation equations (1),(2), (3), (4), (5), (8), (9), (23), and (24), nine equations which are discussed in previous sections must be solved for the nine dependent variables: the gas-phase volume fraction ε_g, the particle-phase volume fraction ε_s, the gas pressure P_g, the gas velocity components u_g and v_g and the solids velocity components u_s and v_s in x-direction and y-direction, respectively,the gas temperature T_g and particle temperature T_s.We need appropriate boundary and initial conditions for the dependent variables listed above to solve the system of equations.

4.1. Boundary conditions

In this section the boundary conditions for the above governing equations, which relate to two dimensional fluidized bed with width "L" and height "H" to allow bed expansion typically i.e. the height of the bed is enough to prevent the particles being ejected from the bed. Boundary conditions are imposed as follow:

$$x=0 \ :v_g=v_s=u_g=u_s=0, \quad \frac{\partial \varepsilon_g}{\partial x}=\frac{\partial T_g}{\partial x}=\frac{\partial T_s}{\partial x}=0$$

$$x=\frac{L}{2} \ :\frac{\partial \varepsilon_g}{\partial x}=\frac{\partial T_g}{\partial x}=\frac{\partial T_s}{\partial x}=\frac{\partial v_g}{\partial x}=\frac{\partial u_g}{\partial x}=\frac{\partial v_s}{\partial x}=\frac{\partial u_s}{\partial x}=0 \ (symmetric)$$

$$y=0 \ :v_g=V_{g,in}, \quad u_g=u_s=v_s=0, \quad \varepsilon_g=1, \quad P_g=P_{g,in}, \quad T_g=T_{g,in}, \quad \frac{\partial T_s}{\partial y}=0$$

$$y=H:\frac{\partial v_g}{\partial y}=\frac{\partial u_g}{\partial y}=\frac{\partial T_g}{\partial y}=0, \quad \varepsilon_g=1, \quad v_s=u_s=0, \quad P_g=P_{atm}$$

4.2. Initial conditions

For setting the initial conditions, the model is divided into two regions: the bed and the freeboard. For each of the regions specified above, an initial condition is specified.

Bed region

$$\varepsilon_g=\varepsilon_{g,mf}, \quad v_g=\frac{V_{g,mf}}{\varepsilon_{g,mf}}, \quad u_g=v_s=u_s=0, \quad T_g=T_{g,in}, \quad T_s=T_{s,in}$$

Freeboard region

$$\varepsilon_g=1, \quad v_g=V_{g,mf}, \quad u_g=v_s=u_s=0, \quad T_g=T_{g,in}$$

5. Finite difference approximation scheme

The conservation equations are transformed into difference equations by using a finite difference scheme.

5.1. Discretization of continuity equations

5.1.1. Particle phase continuity equation

The particle phase continuity equation, Eq.(1), is discretized at the node i; j in an explicit form as:

$$\begin{aligned}(\varepsilon_s)_{i,j}^{n+1}=(\varepsilon_s)_{i,j}^{n}&-\frac{\Delta t}{\Delta x}(u_s)_{i,j}^{n}\begin{cases}(\varepsilon_s)_{i,j}^{n}-(\varepsilon_s)_{i-1,j}^{n} & ,if\,(u_s)_{i,j}^{n}\geq 0.0\\(\varepsilon_s)_{i+1,j}^{n}-(\varepsilon_s)_{i,j}^{n} & ,if\,(u_s)_{i,j}^{n}<0.0\end{cases}\\&-\frac{\Delta t}{\Delta y}(v_s)_{i,j}^{n}\begin{cases}(\varepsilon_s)_{i,j}^{n}-(\varepsilon_s)_{i,j-1}^{n} & ,if\,(v_s)_{i,j}^{n}\geq 0.0\\(\varepsilon_s)_{i,j+1}^{n}-(\varepsilon_s)_{i,j}^{n} & ,if\,(v_s)_{i,j}^{n}<0.0\end{cases}\end{aligned} \tag{40}$$

The gas phase volume fraction is then calculated explicitly as:

http://dx.doi.org/10.5772/52072

$$\left(\varepsilon_g\right)_{i,j}^{n+1} = 1 - \left(\varepsilon_s\right)_{i,j}^{n+1} \tag{41}$$

5.1.2. Gas phase continuity equation

The gas continuity equation residual, d_g, is discretized at i; j in a fully implicit way:

$$d_g = \left(\varepsilon_g\right)_{i,j}^{n+1} - \left(\varepsilon_g\right)_{i,j}^{n} + \frac{\Delta t}{\Delta x}\left(u_g\right)_{i,j}^{n+1}\begin{cases}\left(\varepsilon_g\right)_{i,j}^{n+1} - \left(\varepsilon_g\right)_{i-1,j}^{n+1} & ,if\left(u_g\right)_{i,j}^{n+1} \geq 0.0 \\ \left(\varepsilon_g\right)_{i+1,j}^{n+1} - \left(\varepsilon_g\right)_{i,j}^{n+1} & ,if\left(u_g\right)_{i,j}^{n+1} < 0.0\end{cases}$$
$$+\frac{\Delta t}{\Delta y}\left(v_g\right)_{i,j}^{n+1}\begin{cases}\left(\varepsilon_g\right)_{i,j}^{n+1} - \left(\varepsilon_g\right)_{i,j-1}^{n+1} & ,if\left(v_g\right)_{i,j}^{n+1} \geq 0.0 \\ \left(\varepsilon_g\right)_{i,j+1}^{n+1} - \left(\varepsilon_g\right)_{i,j}^{n+1} & ,if\left(v_g\right)_{i,j}^{n+1} < 0.0\end{cases} \tag{42}$$

5.2. Discretization of combined momentum equations

The combined momentum equations may after a time discretization be expressed in the following forms for x and y directions:

x –direction:

$$\rho_s\left[\frac{\left(u_s\right)_{i,j}^{n+1} - \left(u_s\right)_{i,j}^{n}}{\Delta t} + \frac{\left(u_s\right)_{i,j}^{n}}{\Delta x}\begin{cases}\left(u_s\right)_{i,j}^{n} - \left(u_s\right)_{i-1,j}^{n} & ,if\left(u_s\right)_{i,j}^{n} \geq 0.0 \\ \left(u_s\right)_{i+1,j}^{n} - \left(u_s\right)_{i,j}^{n} & ,if\left(u_s\right)_{i,j}^{n} < 0.0\end{cases}\right]$$
$$+\rho_s\left[\frac{\left(v_s\right)_{i,j}^{n}}{\Delta y}\begin{cases}\left(u_s\right)_{i,j}^{n} - \left(u_s\right)_{i,j-1}^{n} & ,if\left(v_s\right)_{i,j}^{n} \geq 0.0 \\ \left(u_s\right)_{i,j+1}^{n} - \left(u_s\right)_{i,j}^{n} & ,if\left(v_s\right)_{i,j}^{n} < 0.0\end{cases}\right]$$
$$+\rho_g\left[\frac{\left(u_g\right)_{i,j}^{n+1} - \left(u_g\right)_{i,j}^{n}}{\Delta t} + \frac{\left(u_g\right)_{i,j}^{n}}{\Delta x}\begin{cases}\left(u_g\right)_{i,j}^{n} - \left(u_g\right)_{i-1,j}^{n} & ,if\left(u_g\right)_{i,j}^{n} \geq 0.0 \\ \left(u_g\right)_{i+1,j}^{n} - \left(u_g\right)_{i,j}^{n} & ,if\left(u_g\right)_{i,j}^{n} < 0.0\end{cases}\right] \tag{43}$$
$$+\rho_g\left[\frac{\left(v_g\right)_{i,j}^{n}}{\Delta y}\begin{cases}\left(u_g\right)_{i,j}^{n} - \left(u_g\right)_{i,j-1}^{n} & ,if\left(v_g\right)_{i,j}^{n} \geq 0.0 \\ \left(u_g\right)_{i,j+1}^{n} - \left(u_g\right)_{i,j}^{n} & ,if\left(v_g\right)_{i,j}^{n} < 0.0\end{cases}\right] = 0$$

y –direction:

$$\rho_s\left[\frac{(v_s)_{i,j}^{n+1}-(v_s)_{i,j}^{n}}{\Delta t}+\frac{(u_s)_{i,j}^{n}}{\Delta x}\begin{cases}(v_s)_{i,j}^{n}-(v_s)_{i-1,j}^{n} & ,if\,(u_s)_{i,j}^{n}\geq 0.0\\ (v_s)_{i+1,j}^{n}-(v_s)_{i,j}^{n} & ,if\,(u_s)_{i,j}^{n}<0.0\end{cases}\right]$$
$$+\rho_s\left[\frac{(v_s)_{i,j}^{n}}{\Delta y}\begin{cases}(v_s)_{i,j}^{n}-(v_s)_{i,j-1}^{n} & ,if\,(v_s)_{i,j}^{n}\geq 0.0\\ (v_s)_{i,j+1}^{n}-(v_s)_{i,j}^{n} & ,if\,(v_s)_{i,j}^{n}<0.0\end{cases}\right]$$
$$-\rho_g\left[\frac{(v_g)_{i,j}^{n+1}-(v_g)_{i,j}^{n}}{\Delta t}+\frac{(u_g)_{i,j}^{n}}{\Delta x}\begin{cases}(v_g)_{i,j}^{n}-(v_g)_{i-1,j}^{n} & ,if\,(u_g)_{i,j}^{n}\geq 0.0\\ (v_g)_{i+1,j}^{n}-(v_g)_{i,j}^{n} & ,if\,(u_g)_{i,j}^{n}<0.0\end{cases}\right] \tag{44}$$
$$-\rho_g\left[\frac{(v_g)_{i,j}^{n}}{\Delta y}\begin{cases}(v_g)_{i,j}^{n}-(v_g)_{i,j-1}^{n} & ,if\,(v_g)_{i,j}^{n}\geq 0.0\\ (v_g)_{i,j+1}^{n}-(v_g)_{i,j}^{n} & ,if\,(v_g)_{i,j}^{n}<0.0\end{cases}\right]=F_y$$

where:

$$F_y=(C_d)_{i,j}^{n}\frac{3\rho_g\left((v_g)_{i,j}^{n+1}-(v_s)_{i,j}^{n+1}\right)\left|(v_g)_{i,j}^{n}-(v_s)_{i,j}^{n}\right|}{4d_p}\left(1-(\varepsilon_s)_{i,j}^{n}\right)^{-1.8}$$
$$-\frac{(\varepsilon_s)_{i,j}^{n}-(\varepsilon_s)_{i,j-1}^{n}}{\Delta y}\left(3.2gd_p(\rho_s-\rho_g)\right) \tag{45}$$
$$+(C_d)_{i,j}^{n}\frac{3\rho_g(\varepsilon_s)_{i,j}^{n}\left((v_g)_{i,j}^{n+1}-(v_s)_{i,j}^{n+1}\right)\left|(v_g)_{i,j}^{n}-(v_s)_{i,j}^{n}\right|}{4d_p}\left(1-(\varepsilon_s)_{i,j}^{n}\right)^{-2.8}$$

5.3. Discretization of energy equations

Particle phase:

$$
\begin{aligned}
&\rho_s C_{p,s}\frac{\left(\varepsilon_s T_s\right)_{i,j}^{n+1}-\left(\varepsilon_s T_s\right)_{i,j}^{n}}{\Delta t}\\
&+\frac{\rho_s C_{p,s}\left(u_s\right)_{i,j}^{n}}{\Delta x}\begin{cases}\left(\varepsilon_s T_s\right)_{i,j}^{n}-\left(\varepsilon_s T_s\right)_{i-1,j}^{n} & ,if\left(u_s\right)_{i,j}^{n}\geq 0.0\\ \left(\varepsilon_s T_s\right)_{i+1,j}^{n}-\left(\varepsilon_s T_s\right)_{i,j}^{n} & ,if\left(u_s\right)_{i,j}^{n}<0.0\end{cases}\\
&+\frac{\rho_s C_{p,s}\left(v_s\right)_{i,j}^{n}}{\Delta y}\begin{cases}\left(\varepsilon_s T_s\right)_{i,j}^{n}-\left(\varepsilon_s T_s\right)_{i,j-1}^{n} & ,if\left(v_s\right)_{i,j}^{n}\geq 0.0\\ \left(\varepsilon_s T_s\right)_{i,j+1}^{n}-\left(\varepsilon_s T_s\right)_{i,j}^{n} & ,if\left(v_s\right)_{i,j}^{n}<0.0\end{cases}\\
&=\frac{\left(\varepsilon_s K_s T_s\right)_{i+1,j}^{n}-2\left(\varepsilon_s K_s T_s\right)_{i,j}^{n}+\left(\varepsilon_s K_s T_s\right)_{i-1,j}^{n}}{\Delta x^2}\\
&+\frac{\left(\varepsilon_s K_s T_s\right)_{i,j+1}^{n}-2\left(\varepsilon_s K_s T_s\right)_{i,j}^{n}+\left(\varepsilon_s K_s T_s\right)_{i,j-1}^{n}}{\Delta y^2}\\
&+\left(h_v\right)_{i,j}^{n+1}\left(\left(T_g\right)_{i,j}^{n+1}-\left(T_s\right)_{i,j}^{n+1}\right)+\left(\varepsilon_s\dot{q}\right)_{i,j}^{n}
\end{aligned}
\tag{46}
$$

Gas Phase:

$$
\begin{aligned}
&\rho_g C_{p,g}\frac{\left(\varepsilon_g T_g\right)_{i,j}^{n+1}-\left(\varepsilon_g T_g\right)_{i,j}^{n}}{\Delta t}\\
&+\frac{\rho_g C_{p,g}\left(u_g\right)_{i,j}^{n}}{\Delta x}\begin{cases}\left(\varepsilon_g T_g\right)_{i,j}^{n}-\left(\varepsilon_g T_g\right)_{i-1,j}^{n} & ,if\left(u_g\right)_{i,j}^{n}\geq 0.0\\ \left(\varepsilon_g T_g\right)_{i+1,j}^{n}-\left(\varepsilon_g T_g\right)_{i,j}^{n} & ,if\left(u_g\right)_{i,j}^{n}<0.0\end{cases}\\
&+\frac{\rho_s C_{p,g}\left(v_g\right)_{i,j}^{n}}{\Delta y}\begin{cases}\left(\varepsilon_g T_g\right)_{i,j}^{n}-\left(\varepsilon_g T_g\right)_{i,j-1}^{n} & ,if\left(v_g\right)_{i,j}^{n}\geq 0.0\\ \left(\varepsilon_g T_g\right)_{i,j+1}^{n}-\left(\varepsilon_g T_g\right)_{i,j}^{n} & ,if\left(v_g\right)_{i,j}^{n}<0.0\end{cases}\\
&=\frac{\left(\varepsilon_g K_g T_g\right)_{i+1,j}^{n}-2\left(\varepsilon_g K_g T_g\right)_{i,j}^{n}+\left(\varepsilon_g K_g T_g\right)_{i-1,j}^{n}}{\Delta x^2}\\
&+\frac{\left(\varepsilon_g K_g T_g\right)_{i,j+1}^{n}-2\left(\varepsilon_g K_g T_g\right)_{i,j}^{n}+\left(\varepsilon_g K_g T_g\right)_{i,j-1}^{n}}{\Delta y^2}\\
&+\left(h_v\right)_{i,j}^{n+1}\left(\left(T_s\right)_{i,j}^{n+1}-\left(T_g\right)_{i,j}^{n+1}\right)
\end{aligned}
\tag{47}
$$

6. Dimensionless numbers

In this section we define a group of four dimensionless numbers which mainly affect and control fluidization field.

6.1. Archimedes number" Ar"

Archimedes Number is used in characterization of the fluidized state and is defined as follow:

$$Ar = \frac{g d_p^3 \rho_g \left(\rho_s - \rho_g\right)}{\mu_g^2} \tag{48}$$

6.2. Density number "De"

Which define as the density ratio:

$$De = \frac{\rho_g}{\rho_s} \tag{49}$$

6.3. Flow number "fl"

Which is defined as:

$$fl = \frac{V_{g,in}}{u_t} \tag{50}$$

6.4. Dimensionless gas velocity "$\Omega^{1/3}$"

Which is defined as:

$$\Omega^{1/3} = \left[\frac{\rho_g^2}{\mu_g g\left(\rho_s - \rho_g\right)}\right]^{1/3} V_{g,in} \tag{51}$$

6.5. Dimensionless time "τ"

Which is defined as:

$$\tau = \frac{t u_t}{d_p} \tag{52}$$

7. Methodology of solution

The method of solution used in the present work is described in details in this section. The procedures of solution are as following:

1. Input the following parameters :

- Total time of calculation
- Total nodes number in x and y direction
- Total height of the bed and free board
- Width of fluidized bed
- Initial height of the bed
- Properties of gas phase ($\mu_g, \varrho_g, k_g, C_{p,g}$)
- Properties of solid particles phase ($\varrho_s, k_s, C_{p,s}$)
- Entering gas velocity
- Acceleration of gravity
- Particle diameter
- Entering gas temperature
- Initial particles temperature

2. Calculate the minimum fluidized velocity from the following equation [6] :

$$V_{g,mf} = 33.7\left[\left(1+3.59X10^{-5}Ar\right)^{0.5}-1\right]\frac{\mu_g}{\left(d_p\rho_g\right)} \tag{53}$$

3. Calculate the gas void fraction at minimum fluidized velocity from the following equation [7]:

$$\in_{g,mf}=\frac{1}{2.1}[0.4+(\frac{200\mu_g V_{g,mf}}{d_p^2(\rho_s-\rho_g)g})^{1/3}] \tag{54}$$

4. Calculate the gas void fraction at entering gas velocity which given by:

$$\in_{g,in}=\frac{1}{2.1}[0.4+(\frac{200\mu_g V_{g,in}}{d_p^2(\rho_s-\rho_g)g})^{1/3}] \tag{55}$$

5. Calculate the particle terminal velocity from the following equation [2]:

$$u_t = \left[-3.809 + \left(3.809^2 + 1.832 Ar^{0.5}\right)^{0.5}\right]^2 \frac{\mu_g}{\rho_g d_p} \tag{56}$$

6. Specify stability of fluidized bed.

7. Determine the Δt, Δx and Δy

8. Determine the fluidized bed height at the entering velocity using the equation [5]:

$$H1 = \frac{\left(1 - \varepsilon_{g,mf}\right)}{\left(1 - \varepsilon_{g,in}\right)} H_{mf} \tag{57}$$

9. Specify Initial and boundary conditions.

10. Call subroutine "cont" to solve particles phase continuity equations and evaluate the new time step particles volume fractions (ε_s^{n+1}) consequently evaluate (ε_g^{n+1}) from equation (41).

Note: we use the excess solid volume correction [8] in a special subroutine:

The correction works out as a posteriori redistribution of the particle phase volume fraction in excess in each cell where:

$$\varepsilon_s \le \varepsilon_{s,\max} \tag{58}$$

$$\varepsilon_{s,\max} = 1 - \varepsilon_{g,mf} \tag{59}$$

and if $\varepsilon_s > \varepsilon_{s,\max}$ then:

$$\varepsilon_s^{ex} = \varepsilon_s - \varepsilon_{s,\max} \tag{60}$$

The balance may be expressed in terms of particle volume fraction:

$$\varepsilon_{s,i,j}^{new} = \varepsilon_{s,i,j}^{old} - \varepsilon_{s,i,j}^{ex} + \frac{\varepsilon_{s,i+1,j}^{ex}}{4} + \frac{\varepsilon_{s,i-1,j}^{ex}}{4} + \frac{\varepsilon_{s,i,j+1}^{ex}}{4} + \frac{\varepsilon_{s,i,j-1}^{ex}}{4} \tag{61}$$

Figure (3) shows the correction mechanism:

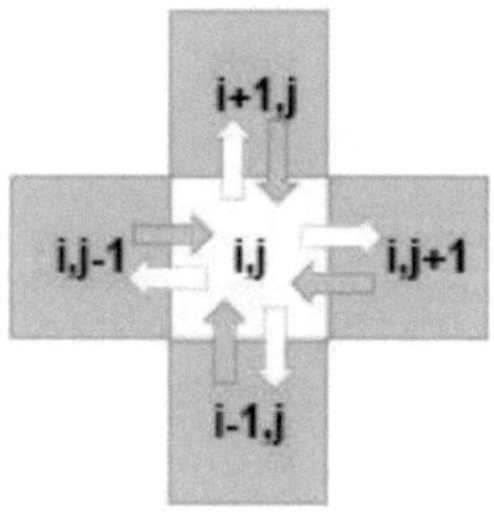

Figure 3. The excess solid volume correction

11. Call subroutine "dirx" to solve momentum equation in x-direction and evaluate at the new time step x-components velocities (u_g^{n+1}*and* u_s^{n+1}). In this subroutine, equations (14) and (43) are solved to get values of x-component new step velocities (u_g^{n+1}*and* u_s^{n+1}). Equation (43) is reduced to the following form :

$$A_{11}(u_s)_{i,j}^{n+1} + A_{12}(u_g)_{i,j}^{n+1} = A_{13} \quad (62)$$

where:

$$A_{11} = \frac{\rho_s}{\Delta t} \quad (63)$$

$$A_{12} = \frac{\rho_g}{\Delta t} \quad (64)$$

$$
\begin{aligned}
A_{13} &= \frac{\rho_s}{\Delta t}\left(u_s\right)^n_{i,j} + \frac{\rho_g}{\Delta t}\left(u_g\right)^n_{i,j} \\
&-\frac{\left(u_s\right)^n_{i,j}}{\Delta x}\begin{cases}\left(u_s\right)^n_{i,j}-\left(u_s\right)^n_{i-1,j} & ,if\left(u_s\right)^n_{i,j}\geq 0.0 \\ \left(u_s\right)^n_{i+1,j}-\left(u_s\right)^n_{i,j} & ,if\left(u_s\right)^n_{i,j}<0.0\end{cases} \\
&-\frac{\left(v_s\right)^n_{i,j}}{\Delta y}\begin{cases}\left(u_s\right)^n_{i,j}-\left(u_s\right)^n_{i,j-1} & ,if\left(v_s\right)^n_{i,j}\geq 0.0 \\ \left(u_s\right)^n_{i,j+1}-\left(u_s\right)^n_{i,j} & ,if\left(v_s\right)^n_{i,j}<0.0\end{cases} \\
&-\frac{\left(u_g\right)^n_{i,j}}{\Delta x}\begin{cases}\left(u_g\right)^n_{i,j}-\left(u_g\right)^n_{i-1,j} & ,if\left(u_g\right)^n_{i,j}\geq 0.0 \\ \left(u_g\right)^n_{i+1,j}-\left(u_g\right)^n_{i,j} & ,if\left(u_g\right)^n_{i,j}<0.0\end{cases} \\
&-\frac{\left(v_g\right)^n_{i,j}}{\Delta y}\begin{cases}\left(u_g\right)^n_{i,j}-\left(u_g\right)^n_{i,j-1} & ,if\left(v_g\right)^n_{i,j}\geq 0.0 \\ \left(u_g\right)^n_{i,j+1}-\left(u_g\right)^n_{i,j} & ,if\left(v_g\right)^n_{i,j}<0.0\end{cases}
\end{aligned}
\quad (65)
$$

so equation (62) and equation (14) result the following system of equations:

$$
\begin{pmatrix} A_{11} & A_{12} \\ (\varepsilon_s)^{n+1}_{i,j} & (\varepsilon_g)^{n+1}_{i,j} \end{pmatrix}\begin{pmatrix}(u_s)^{n+1}_{i,j} \\ (u_g)^{n+1}_{i,j}\end{pmatrix} = \begin{pmatrix} A_{13} \\ U_{gin}\end{pmatrix}
$$

Which are solved to get (u_g^{n+1} *and* u_s^{n+1})

12. Call subroutine "diry" to solve momentum equation in y-direction and evaluate at the new time step y-components velocities (v_g^{n+1} *and* v_s^{n+1}). In this subroutine, equations (13) and (44) are used to get values of (v_g^{n+1} *and* v_s^{n+1}). Equation (44) is reduced to the following form :

$$
B_{11}(v_s)^{n+1}_{i,j} + B_{12}(v_g)^{n+1}_{i,j} = B_{13} \quad (66)
$$

where:

$$
B_{11} = \frac{\rho_s}{\Delta t} + \beta\left(1 + \frac{\left(\varepsilon_s\right)^n_{i,j}}{\left(1-\left(\varepsilon_s\right)^n_{i,j}\right)}\right) \quad (67)
$$

http://dx.doi.org/10.5772/52072

where:

$$\beta = C_d \frac{3(\varepsilon_s)_{i,j}^n \rho_g \left|(v_g)_{i,j}^n - (v_s)_{i,j}^n\right|}{4d_p} (1-(\varepsilon_s)_{i,j}^n)^{-1.8} \tag{68}$$

$$B_{12} = -\frac{\rho_g}{\Delta t} - \beta\left(1 + \frac{(\varepsilon_s)_{i,j}^n}{\left(1-(\varepsilon_s)_{i,j}^n\right)}\right) \tag{69}$$

$$\begin{aligned}
B_{13} = {} & \frac{\rho_s}{\Delta t}(v_s)_{i,j}^n + \frac{\rho_g}{\Delta t}(v_g)_{i,j}^n \\
& - \frac{(u_s)_{i,j}^n}{\Delta x}\begin{cases}(v_s)_{i,j}^n - (v_s)_{i-1,j}^n & ,if\,(u_s)_{i,j}^n \geq 0.0 \\ (v_s)_{i+1,j}^n - (v_s)_{i,j}^n & ,if\,(u_s)_{i,j}^n < 0.0\end{cases} \\
& - \frac{(v_s)_{i,j}^n}{\Delta y}\begin{cases}(v_s)_{i,j}^n - (v_s)_{i,j-1}^n & ,if\,(v_s)_{i,j}^n \geq 0.0 \\ (v_s)_{i,j+1}^n - (v_s)_{i,j}^n & ,if\,(v_s)_{i,j}^n < 0.0\end{cases} \\
& - \frac{(u_g)_{i,j}^n}{\Delta x}\begin{cases}(v_g)_{i,j}^n - (v_g)_{i-1,j}^n & ,if\,(u_g)_{i,j}^n \geq 0.0 \\ (v_g)_{i+1,j}^n - (v_g)_{i,j}^n & ,if\,(u_g)_{i,j}^n < 0.0\end{cases} \\
& - \frac{(v_g)_{i,j}^n}{\Delta y}\begin{cases}(v_g)_{i,j}^n - (v_g)_{i,j-1}^n & ,if\,(v_g)_{i,j}^n \geq 0.0 \\ (v_g)_{i,j+1}^n - (v_g)_{i,j}^n & ,if\,(v_g)_{i,j}^n < 0.0\end{cases} \\
& - \frac{(\varepsilon_s)_{i,j}^n - (\varepsilon_s)_{i,j-1}^n}{dy}\left(3.2 g d_p (\rho_s - \rho_g)\right)
\end{aligned} \tag{70}$$

So equation (66) and equation (13) result the following system of equations:

$$\begin{pmatrix} B_{11} & B_{12} \\ (\varepsilon_s)_{i,j}^{n+1} & (\varepsilon_g)_{i,j}^{n+1} \end{pmatrix} \begin{pmatrix} (v_s)_{i,j}^{n+1} \\ (v_g)_{i,j}^{n+1} \end{pmatrix} = \begin{pmatrix} B_{13} \\ V_{gin} \end{pmatrix}$$

Which are solved to get $(v_g^{n+1} and\ v_s^{n+1})$

13. Call subroutine "temp" to solve energy equation and evaluate the new time step gas and solid particles temperatures $(T_g^{n+1} and\ T_s^{n+1})$.In this subroutine, equations (46) and (47) are used to get values of new time step temperatures for both phases (T_g^{n+1} *and* T_s^{n+1}). Equation (46) is reduced to the following form :

$$C_{11}(T_s)_{i,j}^{n+1} + C_{12}(T_g)_{i,j}^{n+1} = C_{13} \tag{71}$$

where:

$$C_{11} = \frac{\rho_s C_{p,s}}{\Delta t}\left(\varepsilon_s^{n+1}\right) + \left(h_v\right)_{i,j}^{n+1} \tag{72}$$

$$C_{12} = \left(h_v\right)_{i,j}^{n+1} \tag{73}$$

$$\begin{aligned} C_{13} = {} & \frac{\rho_s C_{p,s}}{\Delta t}\left(\varepsilon_s T_s\right)_{i,j}^{n} \\ & -\frac{\rho_s C_{p,s}\left(u_s\right)_{i,j}^{n}}{\Delta x}\begin{cases}\left(\varepsilon_s T_s\right)_{i,j}^{n} - \left(\varepsilon_s T_s\right)_{i-1,j}^{n} & ,if\left(u_s\right)_{i,j}^{n} \geq 0.0 \\ \left(\varepsilon_s T_s\right)_{i+1,j}^{n} - \left(\varepsilon_s T_s\right)_{i,j}^{n} & ,if\left(u_s\right)_{i,j}^{n} < 0.0\end{cases} \\ & -\frac{\rho_s C_{p,s}\left(v_s\right)_{i,j}^{n}}{\Delta y}\begin{cases}\left(\varepsilon_s T_s\right)_{i,j}^{n} - \left(\varepsilon_s T_s\right)_{i,j-1}^{n} & ,if\left(v_s\right)_{i,j}^{n} \geq 0.0 \\ \left(\varepsilon_s T_s\right)_{i,j+1}^{n} - \left(\varepsilon_s T_s\right)_{i,j}^{n} & ,if\left(v_s\right)_{i,j}^{n} < 0.0\end{cases} \\ & +\frac{\left(\varepsilon_s K_s T_s\right)_{i+1,j}^{n} - 2\left(\varepsilon_s K_s T_s\right)_{i,j}^{n} + \left(\varepsilon_s K_s T_s\right)_{i-1,j}^{n}}{\Delta x^2} \\ & +\frac{\left(\varepsilon_s K_s T_s\right)_{i,j+1}^{n} - 2\left(\varepsilon_s K_s T_s\right)_{i,j}^{n} + \left(\varepsilon_s K_s T_s\right)_{i,j-1}^{n}}{\Delta y^2} + (\varepsilon_s \dot{q})_{i,j}^{n} \end{aligned} \tag{74}$$

Also equation (47) is reduced to the following form:

$$D_{11}(T_s)_{i,j}^{n+1} + D_{12}(T_g)_{i,j}^{n+1} = D_{13} \tag{75}$$

where:

$$D_{11} = \left(h_v\right)_{i,j}^{n+1} \tag{76}$$

http://dx.doi.org/10.5772/52072

$$D_{12} = \frac{\rho_g C_{p,g}}{\Delta t}\left(\varepsilon_g^{n+1}\right) + \left(h_v\right)_{i,j}^{n+1} \tag{77}$$

$$\begin{aligned} D_{13} = {} & \rho_g C_{p,g} \frac{\left(\varepsilon_g T_g\right)_{i,j}^{n}}{\Delta t} \\ & - \frac{\rho_g C_{p,g}\left(u_g\right)_{i,j}^{n}}{\Delta x} \begin{cases} \left(\varepsilon_g T_g\right)_{i,j}^{n} - \left(\varepsilon_g T_g\right)_{i-1,j}^{n} & ,if\left(u_g\right)_{i,j}^{n} \geq 0. \\ \left(\varepsilon_g T_g\right)_{i+1,j}^{n} - \left(\varepsilon_g T_g\right)_{i,j}^{n} & ,if\left(u_g\right)_{i,j}^{n} < 0.0 \end{cases} \\ & - \frac{\rho_s C_{p,g}\left(v_g\right)_{i,j}^{n}}{\Delta y} \begin{cases} \left(\varepsilon_g T_g\right)_{i,j}^{n} - \left(\varepsilon_g T_g\right)_{i,j-1}^{n} & ,if\left(v_g\right)_{i,j}^{n} \geq 0.0 \\ \left(\varepsilon_g T_g\right)_{i,j+1}^{n} - \left(\varepsilon_g T_g\right)_{i,j}^{n} & ,if\left(v_g\right)_{i,j}^{n} < 0.0 \end{cases} \\ & + \frac{\left(\varepsilon_g K_g T_g\right)_{i+1,j}^{n} - 2\left(\varepsilon_g K_g T_g\right)_{i,j}^{n} + \left(\varepsilon_g K_g T_g\right)_{i-1,j}^{n}}{\Delta x^2} \\ & + \frac{\left(\varepsilon_g K_g T_g\right)_{i,j+1}^{n} - 2\left(\varepsilon_g K_g T_g\right)_{i,j}^{n} + \left(\varepsilon_g K_g T_g\right)_{i,j-1}^{n}}{\Delta y^2} \end{aligned} \tag{78}$$

so equation (71) and equation (75) result the following system of equations:

$$\begin{pmatrix} C_{11} & C_{12} \\ D_{11} & D_{12} \end{pmatrix} \begin{pmatrix} \left(T_s\right)_{i,j}^{n+1} \\ \left(T_g\right)_{i,j}^{n+1} \end{pmatrix} = \begin{pmatrix} C_{13} \\ D_{13} \end{pmatrix}$$

Which are solved to get (T_g^{n+1} *and* T_s^{n+1})

14. Make gas residual check,which given from the equation (42):

- If $|d_g(i, j)| \leq \delta$ go to step 15, where δ is a small positive value δ=5X10$^{-3.}$
- Else adjust $\varepsilon_g^{n+1}$by :

$$\begin{aligned} \left(\varepsilon_g\right)_{i,j}^{n+1} = \left(\varepsilon_g\right)_{i,j}^{n} - \frac{dt}{\Delta x}\left(u_g\right)_{i,j}^{n+1} & \begin{cases} \left(\varepsilon_g\right)_{i,j}^{n+1} - \left(\varepsilon_g\right)_{i-1,j}^{n+1} & ,if\left(u_g\right)_{i,j}^{n+1} \geq 0.0 \\ \left(\varepsilon_g\right)_{i+1,j}^{n+1} \quad \left(\varepsilon_g\right)_{i,j}^{n+1} & ,if\left(u_g\right)_{i,j}^{n+1} < 0.0 \end{cases} \\ - \frac{\Delta t}{\Delta y}\left(v_g\right)_{i,j}^{n+1} & \begin{cases} \left(\varepsilon_g\right)_{i,j}^{n+1} - \left(\varepsilon_g\right)_{i,j-1}^{n+1} & ,if\left(v_g\right)_{i,j}^{n+1} \geq 0.0 \\ \left(\varepsilon_g\right)_{i,j+1}^{n+1} - \left(\varepsilon_g\right)_{i,j}^{n+1} & ,if\left(v_g\right)_{i,j}^{n+1} < 0.0 \end{cases} \end{aligned} \tag{79}$$

and calculate ε_s^{n+1} from equation (41).

- Go to step 11 and calculate new time step velocities

15. Calculate the gas pressure values.

16. Calculate dimensionless numbers.

17. End program.

8. Parametric study of the hydrodynamic and thermal results

The present work results show the effect of variation of several bed parameters such as particle diameter, terminal velocity of the particle, minimum fluidized velocity of the particle input gas velocity, fluidized material type and heat generation by particles on the hydrodynamic and thermal behavior of fluidized bed. In this section the effect of different parameters in hydrodynamic and thermal performance of fluidized bed is analyzed in detail.

8.1. Effect of particle diameter

Particle diameter is the most influential parameter in the overall fluidized bed performance. In view of that fact, the bed material in a fluidized bed is characterized by a wide range of particle diameter, so that the effect of particle diameter is analyzed in details. In this section particle diameter is changed from 100 μm to 1000 μm for sand as fluidized material to study the effect of particle diameter on the fluidization performance.

One of the important parameter of the fluidized bed study is the total pressure drop across the bed. Although it is constant after beginning of fluidization and equal to the weight of the bed approximately. But its value changes with change of particle diameter.The effect of change of particle diameter on total pressure drop is very important in design and cost of fluidized bed. Figure (4) shows that effect for sand particles of different diameters (100, 200, 300, 400, 500, 600, 700, 800, 900 and 1000 μm). It is clear that the pressure drop increase with increase of particle diameter, which means that small particle is betters in design of fluidized bed cost.

A key parameter in the Thermal analysis of fluidized bed is the average heat transfer coefficient. Figure (5) shows the variation of average heat transfer coefficient with the time for different particles diameter from 100 to 1000 μm sand particles.It is observed from the figure that particles with small size show higher values of average heat transfer coefficient. The particles with diameter 100 μm show higher for average heat transfer coefficient reaches to about 3 times of particles with diameter 1000 μm.

Particles diameter determines the type of particles on Geldart diagram, consequently the behavior of the fluidized bed.

Figures from (6) to (13) illustrate the effect of particles diameter on hydrodynamic and thermal behavior for particle type-B. These figures explain the high disturbance in different bed

http://dx.doi.org/10.5772/52072

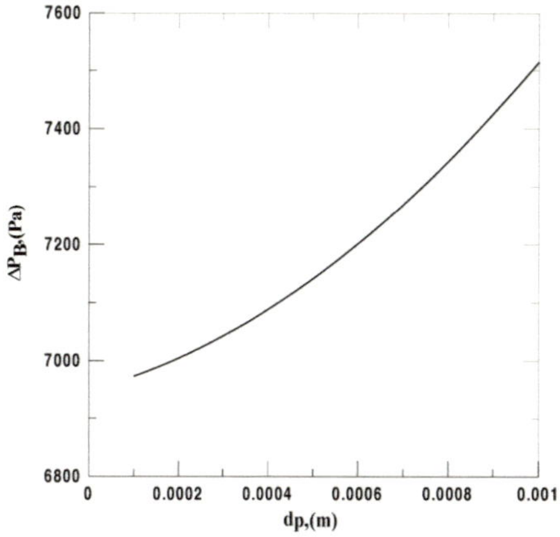

Figure 4. Effect of particle diameter on total pressure drop across fluidized bed

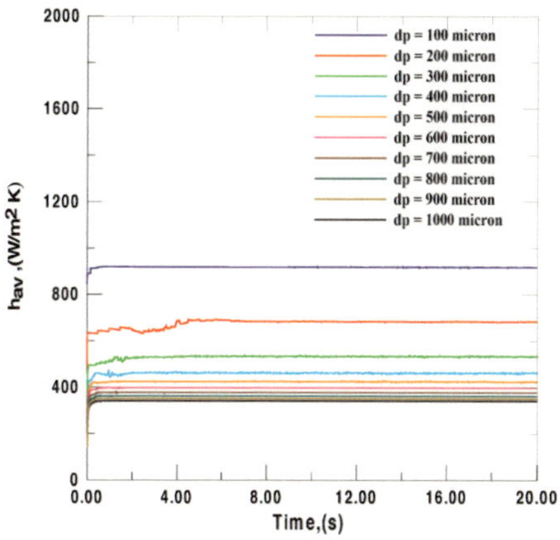

Figure 5. Effect of particle diameter on average heat transfer coefficient in fluidized bed

parameters (ε_g, ϱ_{sus},v_g,v_s,u_g,u_s,T_g,T_s) due to the bubbles formation. The disturbance decreases with increase of particles diameter. This may be due to come near D-type under uniform fluidization.

Fluidized bed is used for wide range of particle diameter. It is better to fluidize particle Dtype in spouted bed to decrease the pressure drop. However, using uniform fluidization for D type give a good and stable thermal behavior of fluidized bed. So in some applications like nuclear reactors the stability and safety is important than cost of pumping power.

Figures from (14) to (21) show the effect of change particle diameter on hydrodynamic and thermal behavior for particle D-type. For this particles type, the behavior of fluidized bed is more uniform in performance than B-type [be consistent with usage and define what the different types are]. Although D-type gives butter fluidization in spouted bed but it gives good performance under uniform fluidization with high pressure drop as shown in figure (4). This means an increase in pumping power and costs to achieve uniform fluidization.

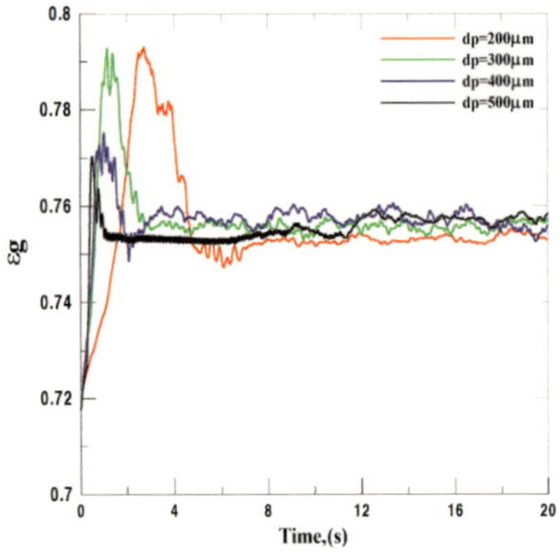

Figure 6. Effect of particle diameter on gas volume fraction (B-type)

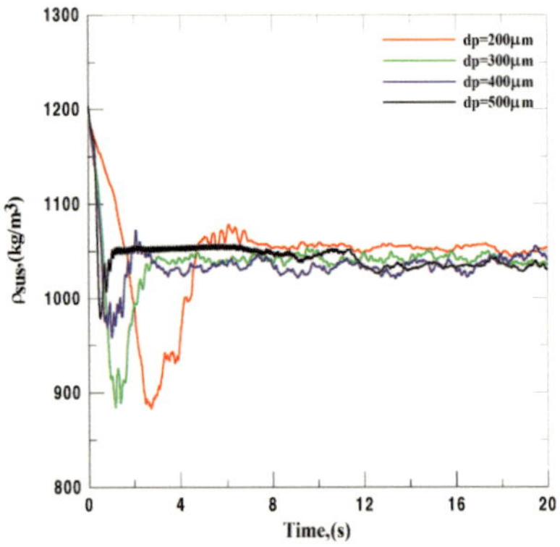

Figure 7. Effect of particle diameter on suspension density (B-type)

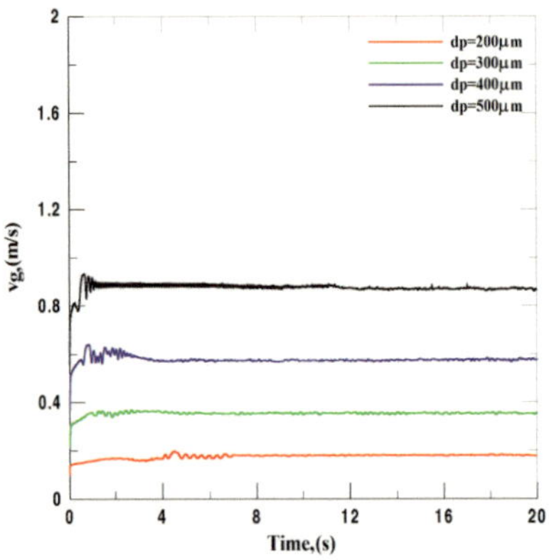

Figure 8. Effect of particle diameter on vertical gas velocity (B-type)

http://dx.doi.org/10.5772/52072

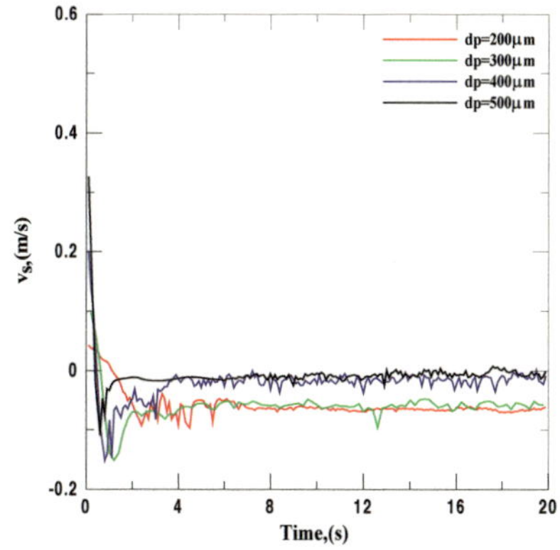

Figure 9. Effect of change particle diameter on vertical particle velocity (B-type)

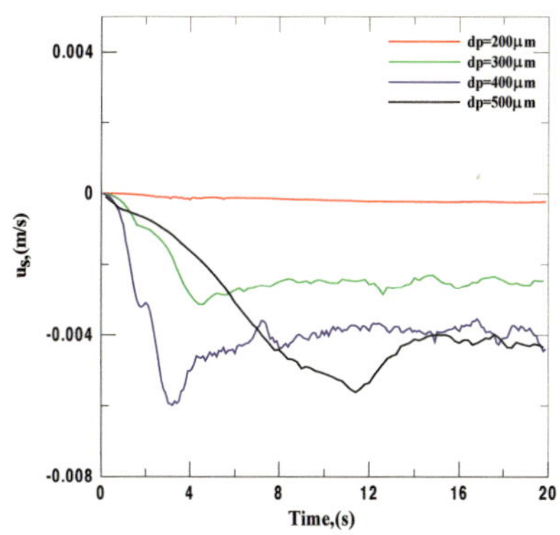

Figure 10. Effect of particle diameter on horizontal particle velocity (B-type)

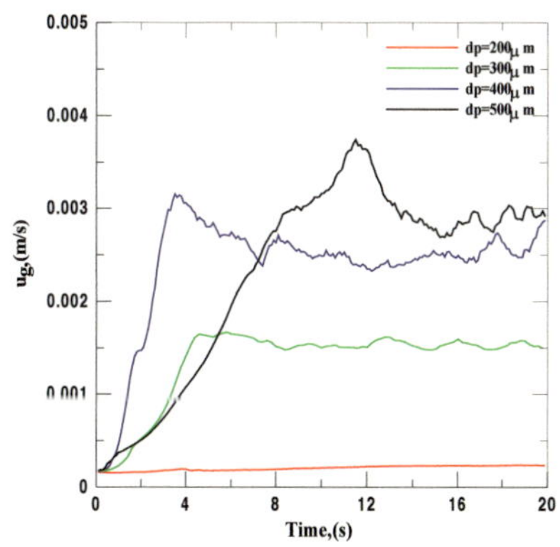

Figure 11. Effect of particle diameter on horizontal gas velocity (B-Type)

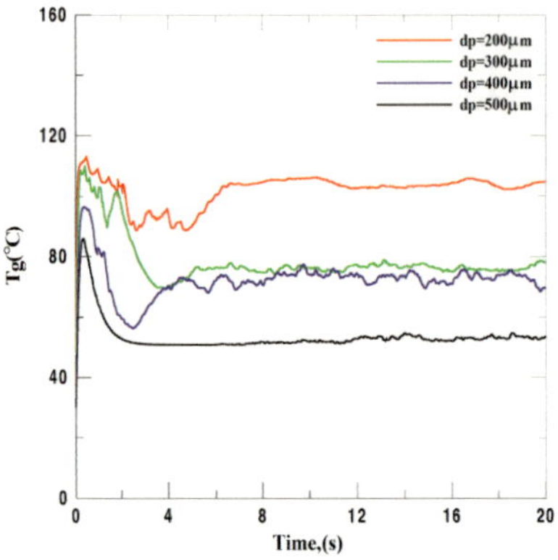

Figure 12. Effect of particle diameter on gas temperature (B-type)

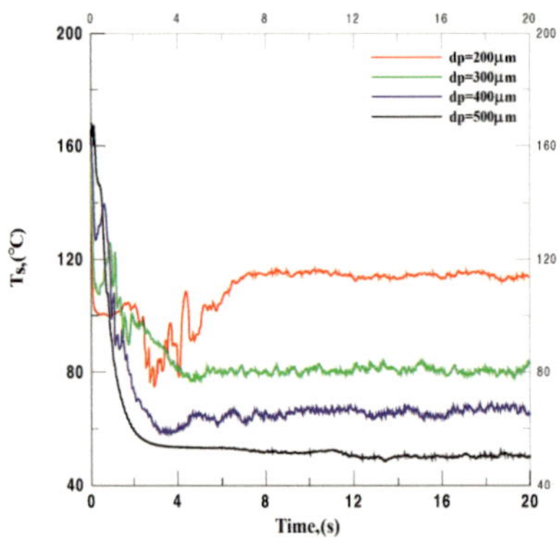

Figure 13. Effect of particle diameter on particle temperature (B-type)

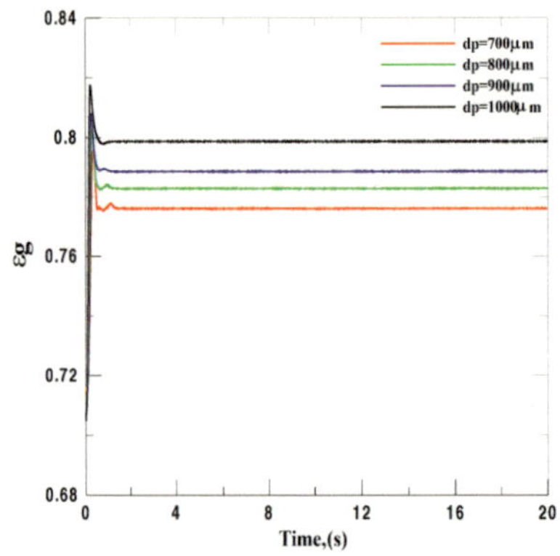

Figure 14. Effect of particle diameter on gas volume fraction (D-type)

http://dx.doi.org/10.5772/52072

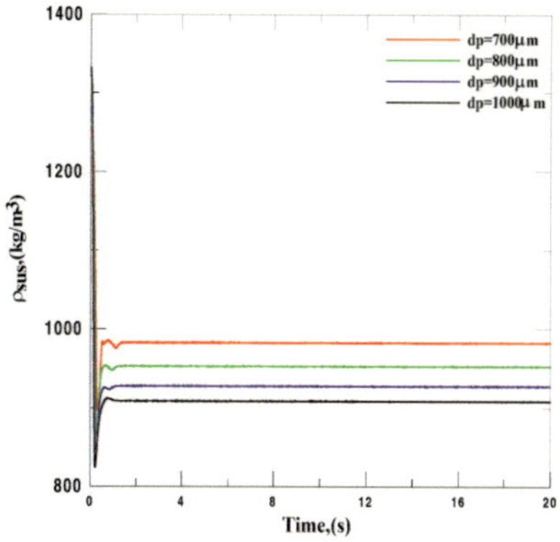

Figure 15. Effect of particle diameter on suspension density (D-type)

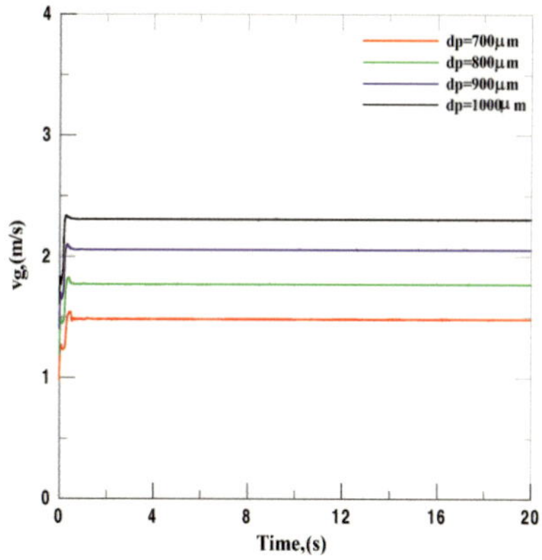

Figure 16. Effect of particle diameter on vertical gas velocity (D-type)

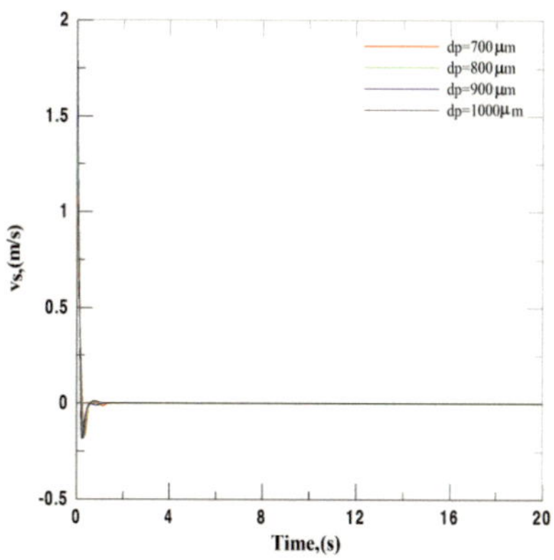

Figure 17. Effect of change particle diameter on vertical particle velocity (D-Type)

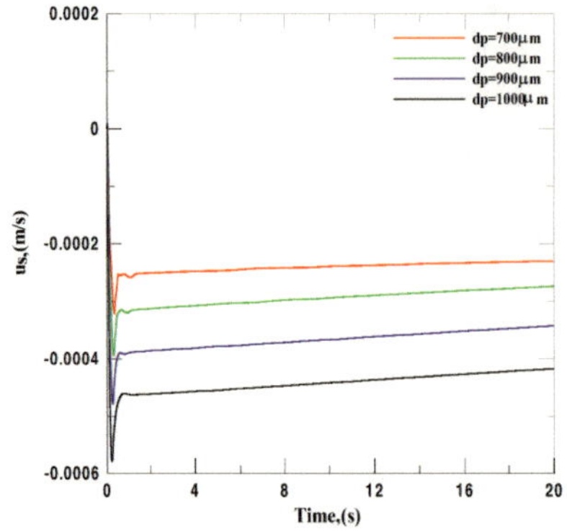

Figure 18. Effect of particle diameter on horizontal particle velocity (D-type)

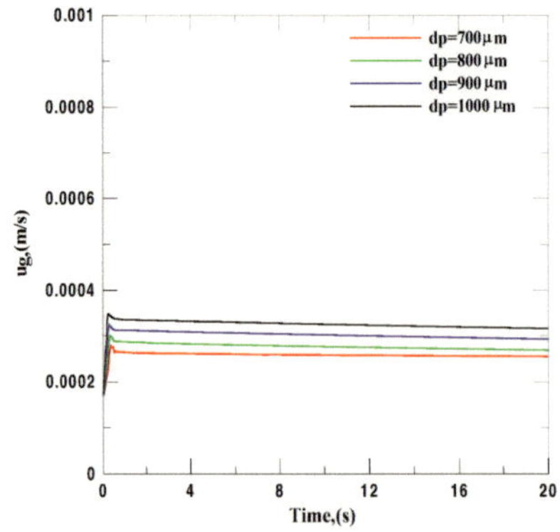

Figure 19. Effect of particle diameter on horizontal gas particle velocity (D-type)

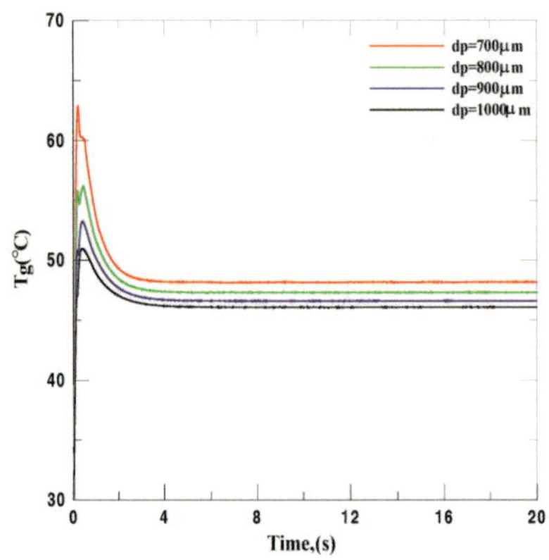

Figure 20. Effect of particle diameter on gas temperature (D-type)

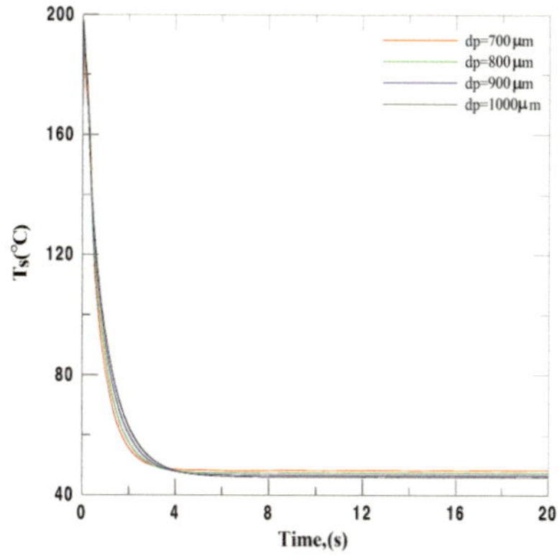

Figure 21. Effect of particle diameter on particle temperature (D-type)

8.2. Effect of input gas velocity

In this section the input gas velocity is changed from one to nine times minimum fluidized velocity for 500 µm sand particles to study the the effect of this parameter on fluidization behavior.

Input gas velocity has an effective role in thermal performance of the fluidized bed. In order to illustrate its effect, the relation between Nusselt number and flow number is described in figure (22). It is clear from this figure that with increase in flow numbers, the Nusselt number increases until reached an optimum flow number where the Nusselt number reaches its maximum value. After this optimum value the increase in flow number is associated with a decrease in Nusselt number. This decrease in Nusselt number may be due to the increase of input gas velocity toward the terminal velocity, consequently the bed goes to be empty bed.

Figure (23) shows the variation of Nusselt number with velocity number. Also the relation between Nusselt number and velocity number has the same trend as the Nusselt number with flow number. This confirms the result from figure (22). This means that there is an optimum input gas velocity to yield the best heat transfer characteristics. This velocity is the target of the fluidized bed designer. The value of this velocity depends on the fluidized gas properties, the fluidized material, particle diameter, and bed geometry.

8.3. Effect of fluidized material type

Type of fluidized material controls hydrodynamic and thermal performance of fluidized bed.It affects on the different parameters of fluidization such as gas volume fraction, suspension density, gas velocity distribution and particle velocity distribution, gas phase temperature and particle phase temperature. In this section different types of materials such as sand, marble, lead, copper, aluminum and steel of particle diameters 1mm are used to study the effect of fluidized materials on the bed performance.

Figure (24) shows the change of gas volume fraction with time for different types of fluidized materials. The figure shows that the gas volume fraction of copper is the highest value

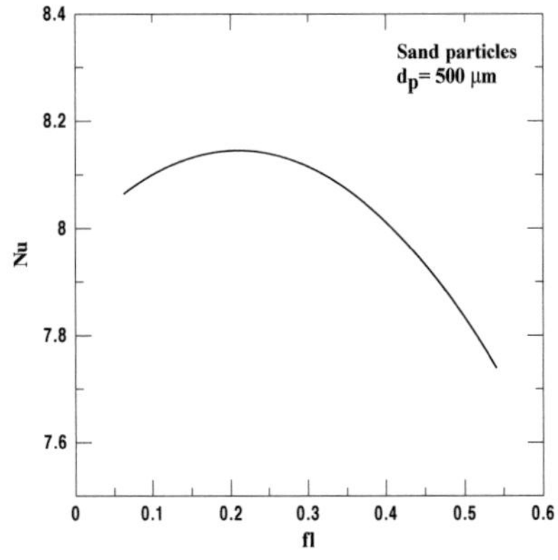

Figure 22. Variation of Nusselt number with flow number

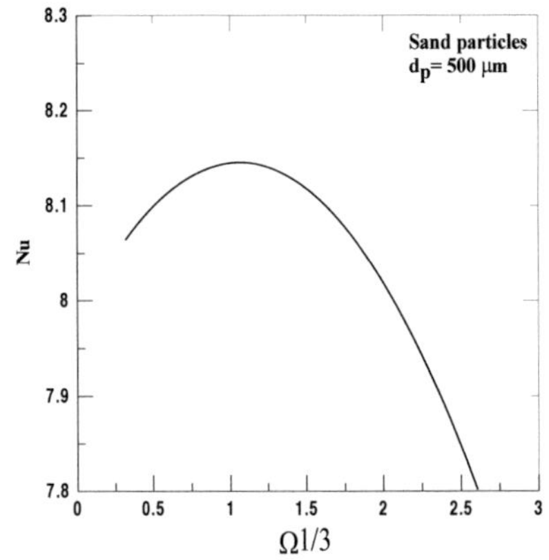

Figure 23. Variation of Nusselt number with velocity number

followed by lead and steel. Gas volume fraction of sand, marble and aluminum are at the same level.

The change of suspension density with time for several types of fluidized materials is shown in Figure (25). It is clear that the highest suspension density lies with the material of the highest density.

Figures (26) to (29) show the effect of change of fluidized material type on horizontal and vertical velocities of gas and particle.

Figures (30) and (40) illustrate the effect of change of fluidized material type on particle and gas temperatures.

http://dx.doi.org/10.5772/52072

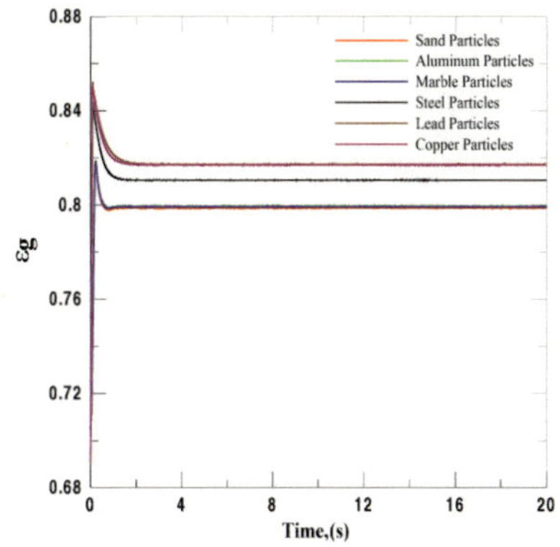

Figure 24. Effect of fluidized material type on gas volume fraction

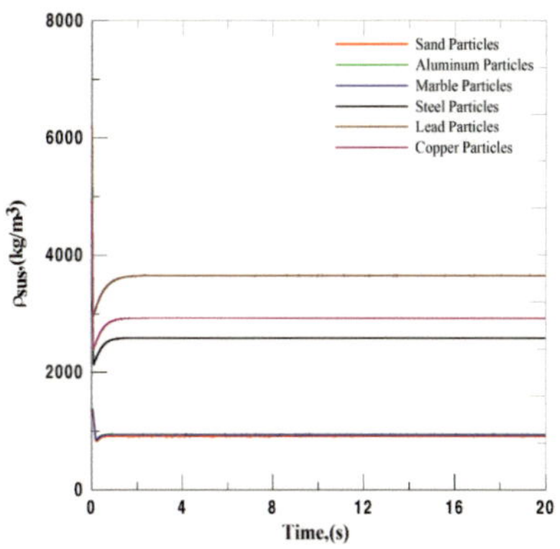

Figure 25. Effect of fluidized material type on suspension density

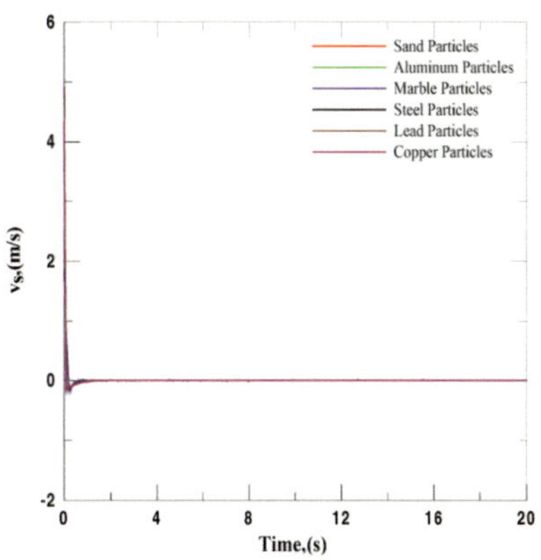

Figure 26. Effect of fluidized material type on vertical particle velocity

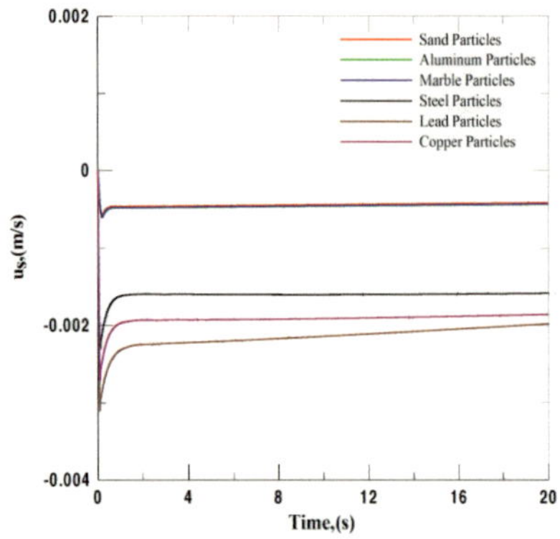

Figure 27. Effect of fluidized material type on horizontal particle velocity

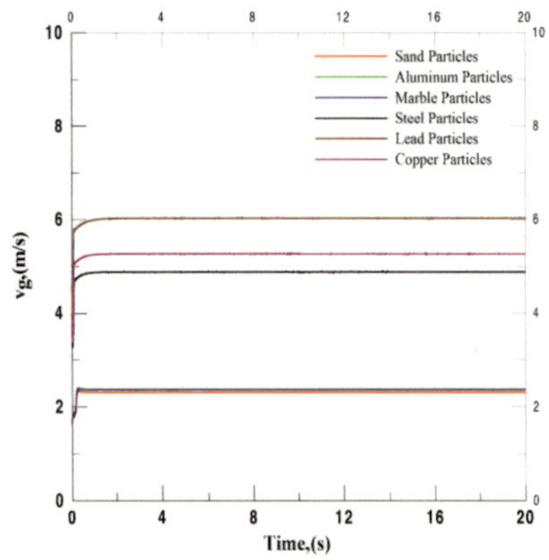

Figure 28. Effect of fluidized material type on vertical gas velocity

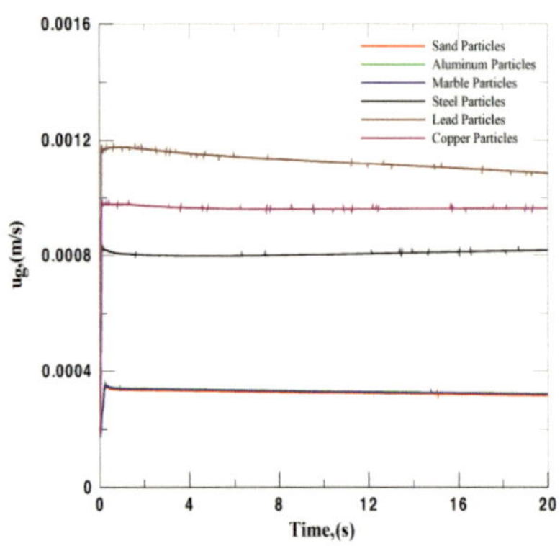

Figure 29. Effect of fluidized material type on horizontal gas velocity

http://dx.doi.org/10.5772/52072

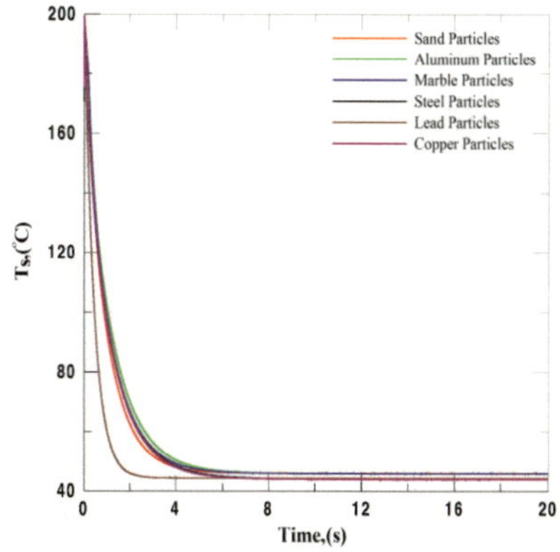

Figure 30. Effect of fluidized material type on particle temperature

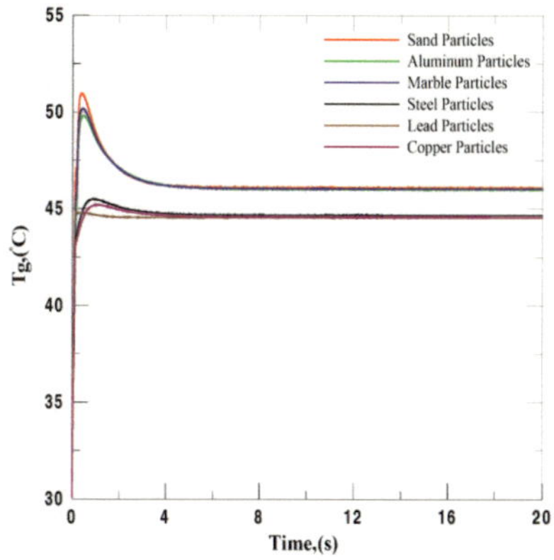

Figure 31. Effect of fluidized material type on gas temperature

8.4. Effect of heat generation by particles

The aluminum particles of 1 mm diameter are fluidized with different value of heat generated in particles (0, 500,1000,1500,2000 and 3000 Watt), the effect of change of heat generated in particles is studied. With the increase of heat generated by particles the gas phase temperature increases as shown in Figure (32). The value of the increase in gas temperature is approximately in range of 3 °C. Figure (33) shows that the particle phase temperature increases with the increase of heat generated by particles. The range of increase is about 45°C. It is clear that the rate of increase in particle temperature is more than the rate of increase in gas temperature, consequently the temperature difference between the two phases increase. Figure (34) illustrates the relation between average heat transfer coefficient and heat generated by particles. The results of the present work shows that the average heat transfer coefficient dos not depend on heat generated by particles and all heat generated by particles converts to temperature difference between the two phases. This result agrees with that of reference [9].

8.5. Terminal velocity effect

Figure (35) shows the relation between terminal velocity and average heat transfer coefficient. It is clear from the figure that with the increase in terminal velocity the average heat transfer coefficient decreases.

8.6. Minimum fluidized velocity effect

Minimum fluidized velocity is the most important parameter in study of fluidization. This velocity distinguishes the fluidized bed from a packed bed and is an indicator that fluidization is occurred. Figure (36) shows the variation of average heat transfer coefficient with minimum fluidized velocity. The average heat transfer coefficient decreases with the increase of minimum fluidized velocity.

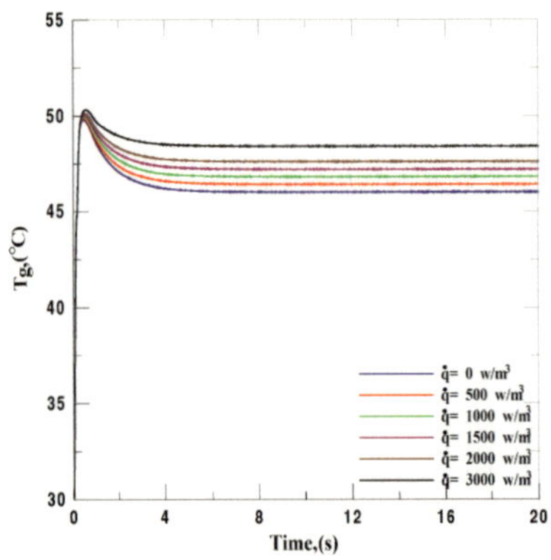

Figure 32. Effect of particle heat generation on gas temperature

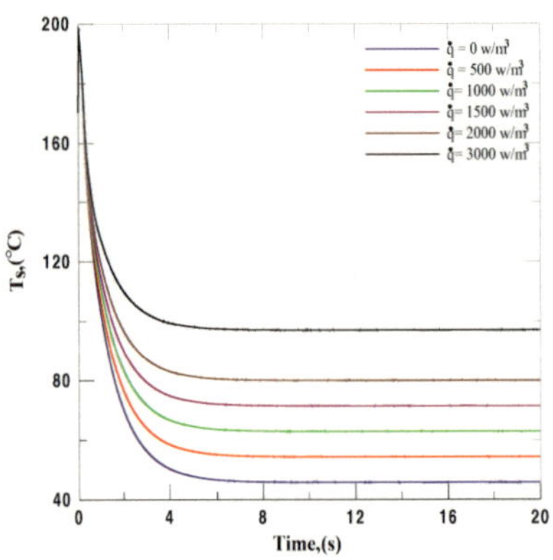

Figure 33. Effect of particle heat generation in particle temperature

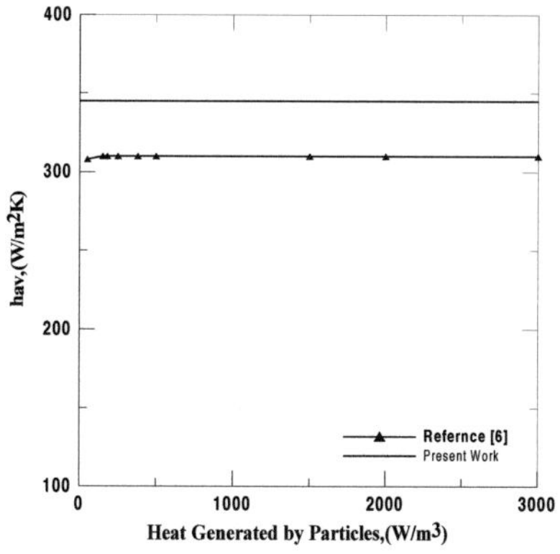

Figure 34. Effect of particle heat generation on average heat transfer coefficient

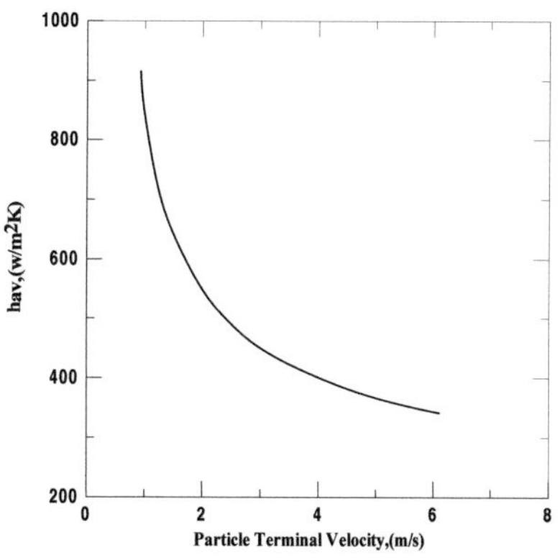

Figure 35. Effect of terminal velocity on average heat transfer coefficient

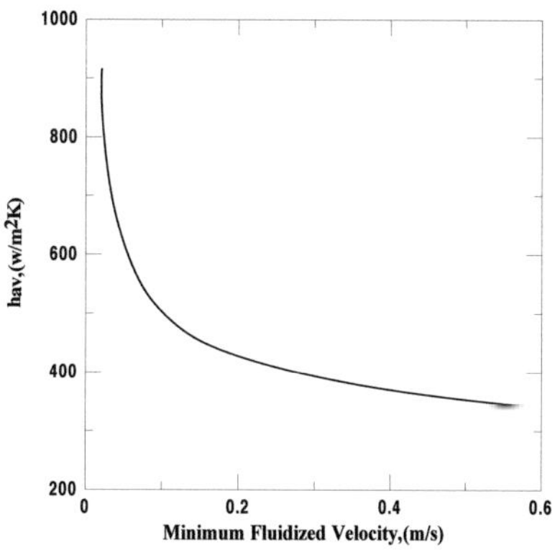

Figure 36. Effect of Minimum Velocity on Average Heat Transfer Coefficient

Nomenclature

Ar	Archimedes number, $Ar = \frac{g d_p^3 \rho_g (\rho_s - \rho_g)}{\mu_g^2}$	
C_d	Drag coefficient, $C_d = \left(0.63 + \frac{4.8}{\mathrm{Re}^{0.5}}\right)^2$	
$C_{p,g}$	Specific heat of fluidizing gas at constant pressure	J/kg.K
$C_{p,s}$	Specific heat of solid particles	J/kg.K
d_g	Residue of the gas continuity equation	Kg/m^3 s
d_p	Mean particle diameter	M
G	Acceleration due to gravity	m/s^2
F_{gx}	Total gas phase force in x direction per unit volume	N /m^3
F_{gy}	Total gas phase force in y direction per unit volume	N /m^3
F_{sx}	Total particle phase force in x direction per unit volume	N /m^3
F_{sy}	Total particle phase force in y direction per unit volume	N /m^3
fl	Flow number , $fl = \frac{V_{g,in}}{u_t}$	
h_{gp}	Heat transfer coefficient between gas phase and particle phase	W/m^2.K
h_v	Volumetric heat transfer coefficient , $h_v = \frac{6(1-\varepsilon_g) h_{gp}}{d_p}$	W/m^3.K
H	Total height of the bed and freeboard	*M*
H_{mf}	Minimum fluidized head of the bed	*M*
H1	Expansion head of bed at the input velocity	*M*
k_g	Thermal conductivity of gas phase	W/m.K
k_s	Thermal conductivity of particle phase	W/m.K
L	Width of the bed	*M*
Nu	Nusselt number based on particle diameter, ($Nu = h_{gp} d_p / k_g$)	
P_g	Gas pressure	Pa
Pr	Prandtl number , $Pr = \mu_g C_{pg} / k_g$	
$\dot{q}$	Rate of heat generated within particle phase	W/m^3
u_g	Gas phase velocity in x direction	m/s
U_{gin}	Input gas velocity to the bed in x direction	m/s
u_s	Particle phase velocity in x direction	m/s
u_t	Particle terminal velocity	m/s
U_r	Relative velocity between two phases	m/s

Re	Reynolds number, $Re = \frac{\varepsilon_g \rho_g d_p \lvert \vec{U}_r \rvert}{\mu_g}$	
S	Stability function	
T	Time	s
T_g	Gas phase temperature	C°
T_s	Particle phase temperature	C°
TFM	Two fluid model	
v_g	gas phase velocity in y direction	m/s
$V_{g,mf}$	Gas minimum fluidized velocity	m/s
$V_{g,in}$	Input gas velocity to the bed in y direction	m/s
v_s	Particle phase velocity in y direction	m/s

Greek Letters

ΔP_B	Total Pressure drop across the bed	Pa
Δt	Time step	s
Δx	Length of cell in the computational grid	m
Δy	height of cell in the computational grid	m
$\in_g$	Gas phase volume fraction	
$\in_{g,mf}$	Gas phase volume fraction at minimum fluidization	
$\in_{in}$	Gas volume fraction at the input velocity	
$\in_s$	Particle phase volume fraction	
ρ_g	Density of gas phase	Kg/m^3
ρ_s	Density of particle phase	Kg/m^3
ρ_{sus}	Suspension density	Kg/m^3
μ_g	Viscosity of gas	Pa.s
δ	Small positive value = $5X10^{-3}$	

Author details

Osama Sayed Abd El Kawi Ali[1,2]

1 Egyptian Nuclear Research Center, Egypt

2 Faculty of Engineering – Al Baha University, Saudi Arabia

References

[1] Farhang Sefidvash 'Fluidized Bed Nuclear Reactor" from internet. (2000). www.regg.ufrgs.br/fbnr_ing.htm,.

[2] John, D. and Jr. Anderson, "Computational Fluid Dynamics",McGraw-Hill,Inc, (1995).

[3] Gibilaro, L. G. Fluidization Dynamics", BUTTER WORTH-HEINEMANN, Oxford, (2001).

[4] Kodikal, J. Nilesh Kodikal and H. Bhavnani Sushil," A Computer Simulation of Hydrodynamics and Heat Transfer at Immersed Surfaces in a Fluidized Bed", Paper Proceedings of the 15th International Conference on Fluidized Bed Combustion, May 16-19, Savannah, Georgia, Copyright by ASME,(1999). (FBC99-0077), 99-0077.

[5] Hans Enwald and Eric PeiranoGemini: A Cartesian Multiblock Finite Difference",Code for Simulation of Gas-Particle Flows", Department of Thermo and Fluid Dynamics, Chalmers University of Technology, 412 96 Goteborg, Sweden,(1997).

[6] Martin RhodesIntroduction to Particle Technology", Published by John Wiley &sons, London, (1998).

[7] Asit KumarDesign, construction,and Operation of 30.5 cm Square Fluidized Bed for Heat Transfer Study", master of technology,Indian institute of technology, Madras,Indian, (1986).

[8] Paola LettieriLuca CammarataGiorgi, D. M. Micale and John Yates, " CFD Simulations of Gas Fluidized Beds Using Alternative Eulerian-Eulerian Modelling Approaches ",International Journal Of Chemical Reactor Engineering, Article 5, (2003)., 1

[9] Osama Sayed Abd ElkawiStudy of Heat Transfer in Fluidized Bed Heat Exchangers", M.Sc. thesis, Mechanical power department, Faculty of engineering, University of Mansoura, Egypt, (2001).